W0111868

# Functional Fillers and Nanoscale Minerals

# Functional Fillers and Nanoscale Minerals

Editor

**Chetna Saraswat**

# Functional Fillers and Nanoscale Minerals

Edited by **Chetna Saraswat**

Printed in 2017

ISBN: 978-1-68117-484-6

Library of Congress Control Number: 2015936596

© 2016 by

SCITUS Academics LLC,
616, Corporate Way, Suite 2, 4766,
Valley Cottage, NY 10989

www.scitusacademics.com

This book contains information obtained from highly regarded resources. Copyright for individual articles remains with the authors as indicated. All chapters are distributed under the terms of the Creative Commons Attribution License, which permits unrestricted use, distribution, and reproduction in any medium, provided the original author and source are credited.

**Notice**

Reasonable efforts have been made to publish reliable data and views articulated in the chapters are those of the individual contributors, and not necessarily those of the editors or publishers. Editors or publishers are not responsible for the accuracy of the information in the published chapters or consequences of their use. The publisher believes no responsibility for any damage or grievance to the persons or property arising out of the use of any materials, instructions, methods or thoughts in the book. The editors and the publisher have attempted to trace the copyright holders of all material reproduced in this publication and apologize to copyright holders if permission has not been obtained. If any copyright holder has not been acknowledged, please write to us so we may rectify.

# Contents

# Preface

Mineral fillers are used in industrial manufacturing processes to extend raw materials and cut costs, and more recently sometimes used for their functionality and other mineral-specific qualities. Mineral additives are widespread in industrial manufacturing processes. Another rapidly emerging technological area is that of nanoscale minerals. This emergence of nanoscale minerals parallels the global pursuit of nanotechnology,:111(1 the use of nanoscale minerals will undoubtedly play an important role in low-cost, high-performance application of nanotechnology. The Functional Fillers and Nanoscale Minerals Symposium was first introduced at the 2003 SME Annual Meeting. The success of the first symposium led to the second symposium of the same name at the 2006 SME Annual Meeting.

**Editor**

# Chapter 1

# Electrical Behaviors of Flame Retardant Huntite and Hydromagnesite Reinforced Polymer Composites

Hüsnügül Yılmaz Atay[1, 2] and Erdal Çelik[1, 2]

[1]Department of Metallurgical and Materials Engineering, Dokuz Eylül University, Tınaztepe Campus, 35160 Izmir, Turkey

[2]Center for Fabrication and Applications of Electronic Materials (EMUM), Dokuz Eylül University, Tınaztepe Campus, 35160 Izmir, Turkey

## ABSTRACT

In our previous work, we studied the physical characteristics (particle size, surface treatment, etc.) of huntite/hydromagnesite mineral in order to be employed as a flame retardant filler. With this respect, electrical properties of the mineral reinforced polymeric composites were investigated in this study. After grinding of huntite/hydromagnesite mineral to the particle size of 10 μm, 1 μm, and 0.1 μm, phase and microstructural analyses were undertaken using X-ray diffraction (XRD) and scanning electron microscopy-energy dispersive spectroscopy (SEM-EDS). The ground minerals with different particle size and content levels were subsequently added to ethylene vinyl acetate copolymer (EVA) to produce composite materials. After fabrication of huntite/hydromagnesite reinforced plastic composite samples, they were characterized by using Fourier transform infrared (FTIR) and SEM-EDS. Electrical properties were measured as a main objective of this paper with Alpha-N high resolution dielectric analyzer as a function of particle size and loading level. Dielectric constant, dissipation factor, specific resistance, and conductivity of the composite materials were measured as a function of frequency. On the other hand, conductivity of Ag-coated and uncoated polymeric composite materials was measured. It was concluded that the electrical properties of plastic composites were improved with reducing the mineral particle size.

# INTRODUCTION

Despite significant advances in synthesis and characterization of polymers, a correct understanding of polymer molecular structure did not emerge until the 1920s. Before then, scientists believed that polymers were clusters of small molecules (called colloids), without definite molecular weights, held together by an unknown force. In 1922, Hermann Staudinger proposed that polymers consisted of long chains of atoms held together by covalent bonds, an idea which did not gain wide acceptance for over a decade and for which Staudinger was ultimately awarded the Nobel Prize [1].

A polymer can be described as macromolecule composed of repeating structural units typically connected by covalent chemical bonds [2]. A large and growing number of commercial polymers are composed of different types of unit attached together by chemical covalent bonds. They are known as copolymers and can comprise just two different units or three and so on. It is one of the common strategies used by molecular engineers to manipulate the properties of polymers to gain just the right combination of properties for a specific application [3].

Due to their low weight and ease of processing, the use of polymers is raised by their remarkable combination of properties in our daily life. They are chosen in preference to many conventional materials. For example typical uses of composites are monocoque structures for aerospace and automobiles, as well as more mundane products like fishing rods and bicycles. The stealth bomber was the first all-composite aircraft, but many passenger aircraft like the Airbus uses an increasing proportion of composites in its fuselage. The quite different physical properties of composites give designers much greater freedom in shaping parts, which is why composite products often look different to conventional products [1].

Even though being used in so many areas and show great facilities, polymers are also known for their relatively high flammability. Beside, most of them are accompanied by corrosive or toxic gases and smoke which are produced while the combustion is continuing [4]. So that it is rising as an important issue to extent polymers' usage for obtaining their fire resisting property for the applications [5]. Hence some ancillary materials are used to make polymers fire resistant. They are added into the compound whose application properties became closely related to the physical properties of the additive itself. Alumina trihydrate, ATH, $(Al_2O_3 \cdot 3H_2O)$, and magnesium hydroxide, MH, $(Mg(OH)_2)$ are the major materials used as fire retardant fillers for polymers [6–8]. These two materials account for more than 50% by weight of the worldwide sales of fire retardants [9,10]. The use of ATH is limited to those polymers processed below about 200°C while MH is stable above 300°C and thus can be used in polymers that must be processed at higher temperatures. Their effectiveness comes from the fact that they both decompose endothermically and consume a large amount of heat, while also liberating water, which can dilute any volatiles and thus decrease the possibility of fire (Equations (1) and (2), [7]). For ATH, decomposition begins near 300°C and consumes 1270 joules per gram (J/g) of ATH; for MH, decomposition begins at somewhat higher temperature, near 400°C, and consumes 1244 J/g of MH [8]. A major use of both ATH and MH is in low smoke, halogen-free wire and cable applications, where there is significant commercial activity [11],

$$Mg_4(CO_3)_3(OH)_2 \cdot 3H_2O \longrightarrow 4MgO + 3CO_2 + 4H_2O \quad (1)$$

$$Mg_3Ca(CO_3)_4 \longrightarrow 3MgO + CaO + 4CO_2. \quad (2)$$

Halogen-containing flame retardants act in the gas phase of a fire. When a polymer burns they decompose by generating free and highly reactive radicals in the gas phase, which are responsible for the propagation of a fire. Then produced incomplete burned substances lead to an increase in smoke density which is more toxic and corrosive [12]. Nonetheless, mineral nonhalogenated flame retardants release less CO and smoke when compared to halogenated flame retardants [6, 8, 13]. Metal hydroxide flame retardant materials are some of commonly used materials in this matter [8]. They firstly decompose endothermically with releasing water. It cools the condensed phase where the fuel is located, slowing its rate of decomposition and pyrolysis. Further, the water released from mineral filler flame retardants does not inhibit the amount of oxygen available for combustion. Instead it dilutes the total amount of fuel available in the pyrolysis stream for combustion [6]. As a result of this action the amount of oxygen, which is able to enter into the flame, is restricted due to the release of water and this avoids the critical fuel/oxygen ratio [14]. Moreover, after degradation, a ceramic: based protective layer was created and this improves insulation giving rise to a smoke-suppressant effect [15]. During combustion, this ceramic-based protective layer plays an important role for the efficient protection of the polymer compound and decreasing the heat release [14]. Besides, this ceramic layer reduces smoke density by adsorbing soot particles [12]. On the other hand, intumescence involves an increase in volume of the burning substrate as a result of network or char formation. For ingressing oxygen to the fuel, this char serves as a barrier and also as a medium in which heat can be dissipated [16]. This char formation eliminates dripping and promoting. As it can be seen from Figure 1 that flame retardant cable jacket formulation at right does not drip, while unfilled sample at left drips [16].

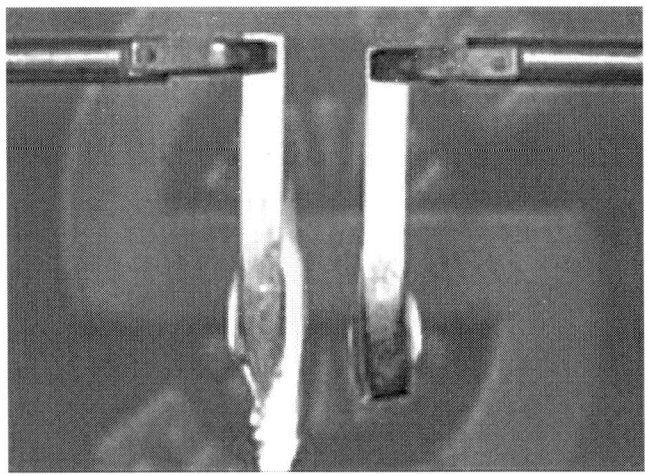

**Figure 1**: Char formation prevents dripping and promoting [16].

In our previous work [17], we studied the physical characteristics (particle size, surface treatment, etc.) of huntite/hydromagnesite mineral in order to be employed as flame retardant filler in vinyl acetate copolymer [6, 7, 10]. It is a halogen-free material with formula $Mg_3Ca(CO_3)_4$ and $Mg_4(CO_3)_3(OH)_2·3H_2O$ [10]. As it decomposes endothermically and evolves water vapor and carbon dioxide, it was obtained a total associated heat of 10.53 J/g and a final weight loss of 56% in the temperature range of 220 and 600°C [17]. Regarding the flame retardancy properties, it was obtained that increasing the loading level of additive [15] and decreasing the size increase the flame retardancy of the polymer composite [18]. This was explained with the fact of increasing surface area. Consequently, it was concluded that HM is a promising flame retardant filler for vinyl acetate copolymers [8].

Although unified by direct concern with the effects produced by electric fields, the subject of the electrical properties polymers covers a diverse range of molecular phenomena [19]. By comparison with metals, where the electrical response is overwhelmingly one of electronic conduction, polymers display a much less striking response. This absence of any overriding conduction does allow, however, a whole set of more subtle electrical effects to be observed more easily. For instance, polarization resulting from distortion and alignment of molecules under the influence of the applied field becomes apparent.

Examination of such polarization not only gives valuable insight into the nature of the electrical response itself but it also provides a powerful means of probing molecular dynamics. For this reason electrical studies form a desirable supplement to studies of purely mechanical properties aimed at reaching an understanding of the behavior of polymers on a molecular basis [20].

# ELECTRICAL MECHANISM OF POLYMERS

Understanding the structure of polymers not only gives a better behavior of chemical resistance but also of the electrical properties. Most polymers are dielectrics or insulators and resist the flow of a current. This is one of the most useful properties of plastics and makes much of our modern society possible through the use of plastics as wire coatings, switches, and other electrical and electronic products. Despite this, dielectric breakdown can occur at sufficiently high voltages to give current transmission and possible mechanical damage to the plastic [21].

The application of a potential difference causes the movement of electrons and when the electrons are free to move there is a flow of current. Metals can be thought of as a collection of atomic nuclei existing in a "sea of electrons" and when a voltage is applied the electrons are free to move and to conduct a current. Polymers and the atoms that make them up have their electrons tightly bound to the central long chain and side groups through "covalent" bonding. Covalent bonding makes it much more difficult for most conventional polymers to support the movement of electrons and therefore they act as insulators [21].

# POLAR AND NONPOLAR POLYMERS

Not all polymers behave the same when subjected to voltage and plastics can be classified as "polar" or "nonpolar" to describe their variations in behavior. The polar plastics do not have a fully covalent bond and there is a slight imbalance in the electronic charge of the molecule. A simple example of this type of behavior would be that of the water molecule ($H_2O$) (see Figure 2). The conventional representation of the molecule is that shown at that right. The two hydrogen atoms are attached to the oxygen atom and the overall molecule has no charge [21].

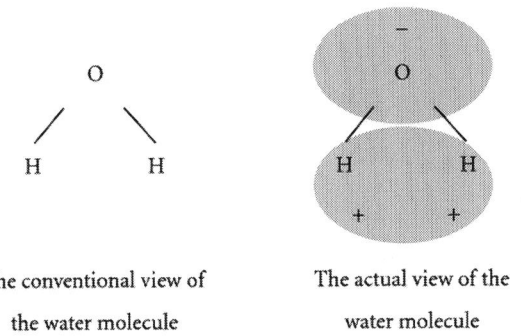

The conventional view of
the water molecule

The actual view of the
water molecule

**Figure 2**: Structure of a water molecule; (a) conventional and (b) actual [21].

In reality, the electrons tend to be around the oxygen atom more than around the hydrogen atoms and this gives the oxygen a slightly negative charge and the hydrogen atoms a slightly positive charge. This is shown in the diagram at right where the grey areas show where the electrons are more often found. The overall water molecule is neutral and does not carry a charge but the imbalance of the electrons creates a "polar" molecule. This "polar dipole" will move in the presence of an electric field and attempt to line up with the electric field in much the same way as a compass needle attempts to line up with the earth's magnetic field.

In polar plastics, dipoles are created by an imbalance in the distribution of electrons and in the presence of an electric field the dipoles will attempt to move to align with the field. This will create "dipole polarization" of the material and because movement of the dipoles is involved there is a time element to the movement. The non-polar plastics are truly covalent and generally have symmetrical molecules. In these materials there are no polar dipoles present and the application of an electric field does not try to align any dipoles. The electric field does, however, move the electrons slightly in the direction of the electric field to create "electron polarization"; in this case the only movement is that of electrons and this is effectively instantaneous. The structure of the polymer determines if it is polar or non-polar and this determines many of the dielectric properties of the plastic [21].

# ELECTRICAL APPLICATIONS OF POLYMERS

The electrical insulating quality inherent in most polymers has long been exploited to constrain and protect currents flowing along chosen paths in conductors and to sustain high electric fields without breaking down. Insulating polymeric materials for early electrical equipment were made from naturally occurring products. For instance, the first trans-Atlantic telephone cables laid in the 1860s were insulated with Gutta-percha, which is one of the polymers extracted from rubber trees. As synthetic high polymers became available in the twentieth century, the range of insulators was continually improved. The great virtue of these new materials, such as polystyrene, was their combination of high quality of insulation with ease of fabrication by molding. Polyethylene, which combines superb insulating properties with moldability and a high degree of toughness and flexibility, arrived on the scene just in time for the more demanding applications of insulation in coaxial cables for radar apparatus and television. More recently extreme requirements for very low-conductivity materials, used in electret microphones, for example, have been met by fluorinated polymers. High-performance thin films have also been developed for various types of capacitors [20].

The choice of material for a particular application naturally depends on being able to reach a good compromise amongst a whole range of considerations, including mechanical properties, ease of fabrication into a final product, and cost. The basic insulating properties of polymers are more than adequate for many purposes, and any development effort may then be primarily concentrated on improving other aspects of the material's performance. High on the list will be a need for good chemical and physical stability in the working environment, which might involve exposure to strong sunlight, organic solvents, and high temperatures. Only on the basis of detailed knowledge and understanding of the molecular structure and behavior of the basic polymers one can hope to approach the optimum in performance. This has motivated the investigation in depth of the electrical behavior of many polymeric systems [20].

Polymers are always good insulators, but that is not to say that a conducting plastic is not desirable. A lightweight, readily moldable, highly conductive material has long been recognized as a worthwhile goal to strive for, and considerable scientific research has been devoted to this. Even though encouraging results have been obtained, there is still a long way to go in developing useful products. Apart from the obvious market for a highly conductive material which could be suitable for power and signal transmission, there is also one for materials having intermediate conductive properties, for example, for flexible heating elements and graded cable insulations. Certain of these can be supplied by modification of existing polymers [20].

As electrical properties of polymers, elastomers, composites, and films are important in a wide range of industries including automotive, aerospace, building products, marine, packaging and consumer goods, and electrical tests, in general, are performed as the measurements of the resistance, conductivity, or charge storage either on the surface or through the plastic composite material. In the light of this a series of composites were prepared using an ethylene vinyl acetate copolymer matrix in the present work. Huntite/hydromagnesite mineral powders were added to ethylene vinyl acetate copolymer at different concentrations to evaluate the electrical properties of the composites.In this sense, properties of complex conductivity, dielectric constant, specific resistance, and dissipation factor measurements were performed to the plastic composite samples.

# EXPERIMENTAL DETAILS

## Preprocessing and Fabrication of Composite Samples

Preprocessing and fabrication of the composites were described in [17] in detail. Briefly, preprocessing includes excavating, crushing, grinding, and separating mineral powders according to their size. Open pit mining technique was used at the quarry to excavate the mineral due to the proximity of deposit to the surface. Then it was comminuted in an impact crusher to approximately 1 cm particle size. In the grinding area, huntite and hydromagnesite mineral was subjected to a high degree of turbulent mixing in cells created between high velocity rotor blades and the high energy process reduces agglomeration of the mineral particles to a minimum.

Huntite/hydromagnesite mineral powders, supplied by Likya Minelco Madencilik (Denizli, Turkey), were subjected to a sedimentation process. At the end of this procedure, three different size products were obtained: 10, 1, and 0.1 µm. As a polymer matrix ethylene-co-vinyl acetate (poly) was used. Polymer composites were a blend of ethylene-co-vinyl acetate and the mineral powders in different ratio from 49% to 69% and size from 0.1 µm to 10 µm. They were processed using a twin screw extruder and subsequently pelletized and compression molded at 160°C to obtain 1 mm thick sheets.

# Characterization

Rigaku D (Max-2200/PC Model XRD) X-ray diffractometer equipment was used to identify the huntite and hydromagnesite phase at 40 kV, 20 mA with a monochromatic CuK irradiation (wavelength, $\lambda$ =0.15418nm) by both -2 mode and 2 scan mode with a scan speed of 8°/min. Fourier transform infrared (FTIR, Perkin Elmer Spectrum BX) absorption spectra of the composite materials were only measured over the range of 4000 to 400 cm$^{-1}$ at room temperature in order to determine organic structure and interaction between polymeric matrix and reinforced material. JEOL JJM 6060 scanning electron microscopy (SEM) was used to examine the microstructural cross-sectional areas of huntite and hydromagnesite reinforced polymeric matrix composite materials.

# Electrical Measurements

Electrical Properties were measured with Alpha-N high resolution dielectric analyzer in detail. By this method, electrical properties of huntite/hydromagnesite reinforced plastic composite samples were measured. Insight into these complex electrical properties of composite materials can be gained from analysis including dielectric constant, conductivity, specific resistance, impedance, capacitance, and dissipation factor.

The dielectric constant is a measure of the influence of a particular dielectric on the capacitance of a condenser. It measures how well a material separates the plates in a capacitor and is defined as the ratio of the capacitance of a set of electrodes with the dielectric material between them to the capacitance of the same electrodes with a vacuum between them. The dielectric constant for a vacuum is 1 and for all other materials it is greater than 1 [21].

Electrical conductivity or specific conductance is a measure of a material's ability to conduct an electric current. To analyse the conductivity of materials exposed to alternating electric fields, it is necessary to treat conductivity as a complex number called the admittivity, complex conductivity. This method is used in applications such as electrical impedance tomography, a type of industrial and medical imaging. Admittivity is the sum of a real component called the conductivity and an imaginary component called the susceptivity [22,23].

Specific electrical resistance is a measure of how strongly a material opposes the flow of electric current. A low resistivity indicates a material that readily allows the movement of electrical charge [22, 23]. Most plastics have very high volume resistivity (in the order of 1016 $\Omega$m) and are therefore good insulators [24].

As for dissipation factor (DF), it is a measure of loss rate of power of a mode of oscillation (mechanical, electrical, or electromechanical) in a dissipative system. It is the reciprocal of quality factor, which represents the quality of oscillation. To illustrate this, electric power is dissipated in all dielectric materials, usually in the form of heat [25].

# RESULTS AND DISCUSSION

## Material Characterization

The XRD pattern of mineral powder (Figure 3) demonstrates that the basic minerals are hydromagnesite ($Mg_4(CO_3)_3(OH)_2 \cdot 3H_2O$), huntite ($Mg_3Ca(CO_3)_4$), and dolomite ($CaMg(CO_3)_2$. As mentioned in the literature [26], morphologically those rocks are composed of variable amounts of siliciclastic and carbonate-clastic debris cemented by dolomite, monohydrocalcite, hydromagnesite, huntite and magnesite. In addition to this, the literature [26, 27] focused on that those minerals are sediments of the lake predominantly composed of gypsum, dolomite, huntite, hydromagnesite, and magnesite.

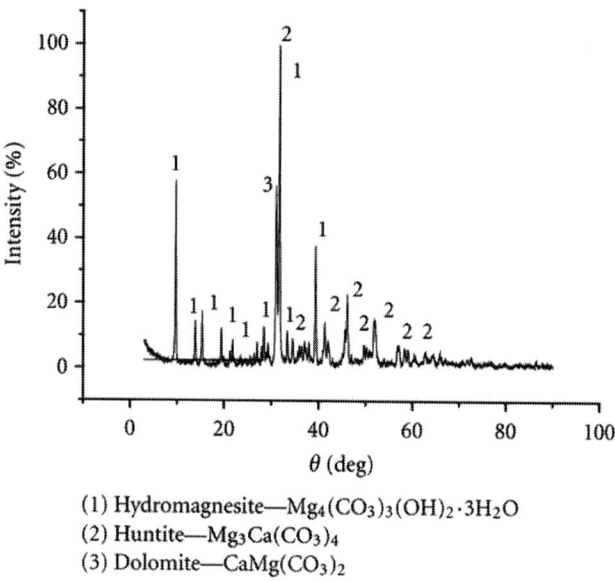

(1) Hydromagnesite—Mg$_4$(CO$_3$)$_3$(OH)$_2$·3H$_2$O
(2) Huntite—Mg$_3$Ca(CO$_3$)$_4$
(3) Dolomite—CaMg(CO$_3$)$_2$

**Figure 3**: XRD pattern of as received huntite and hydromagnesite mineral powder.

Figure 4 elucidates FTIR analysis of huntite/hydromagnesite containing plastic composite materials including different mineral particle sizes. The FTIR spectrum of calcium magnesium carbonate is quite characteristic with a very intense broad band centering at 2750–4000 cm$^{-1}$ there is another sharp band at 3750 cm$^{-1}$. This could be due to O-H groups in the samples. Additionally, lower intensity absorptions at 600–700 cm$^{-1}$ can be observed from magnesium carbonate. In the samples with different size distributions, there are small differences in the results. As FTIR analysis is related with the molecular bonding, getting the material size to nanoscale, the bond numbers and the intensities decreased.

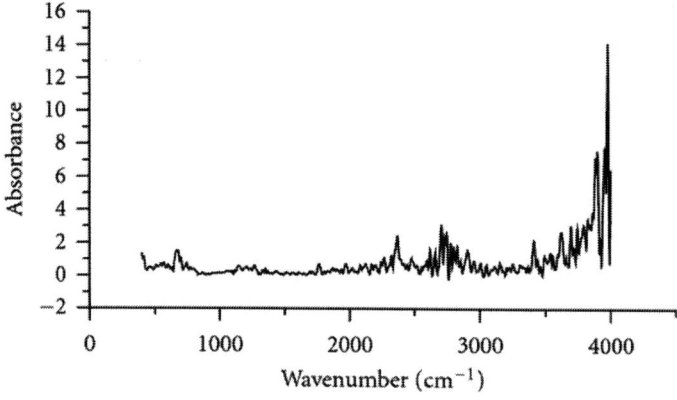

(a)

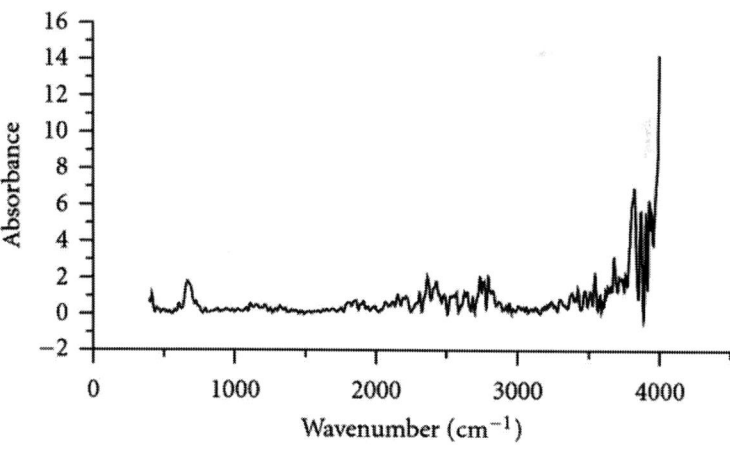

(b)

**Figure 4**: FTIR analysis of huntite and hydromagnesite reinforced plastic composite material including (a) 10 μm, (b) 1 μm, and (c) 0.1 μm particle size.

SEM micrographs of huntite and hydromagnesite reinforced plastic composite material are demonstrated in Figure 5. Morphological features that contribute to electrical properties are particle size and content of huntite and hydromagnesite powders in the plastic matrix. Notice that changing of huntite and hydromagnesite particle size can be clearly seen in the composite material from 10 µm to 0.1 µm (100 nm). It can be pointed out from the figures that by getting nanoscale particle sizes, much denser structures can be produced in the composite material. Besides, it is indicated in the literature [28] that in order to achieve suitable plastic formulations it is necessary to reduce strongly the mineral particle size. This may affect not only the morphology but the crystalline characteristics of the material. Therefore, electrical properties, flame retardancy, and mechanical properties are strongly influenced by virtue of Nano sized huntite and hydromagnesite reinforced materials.

(a)

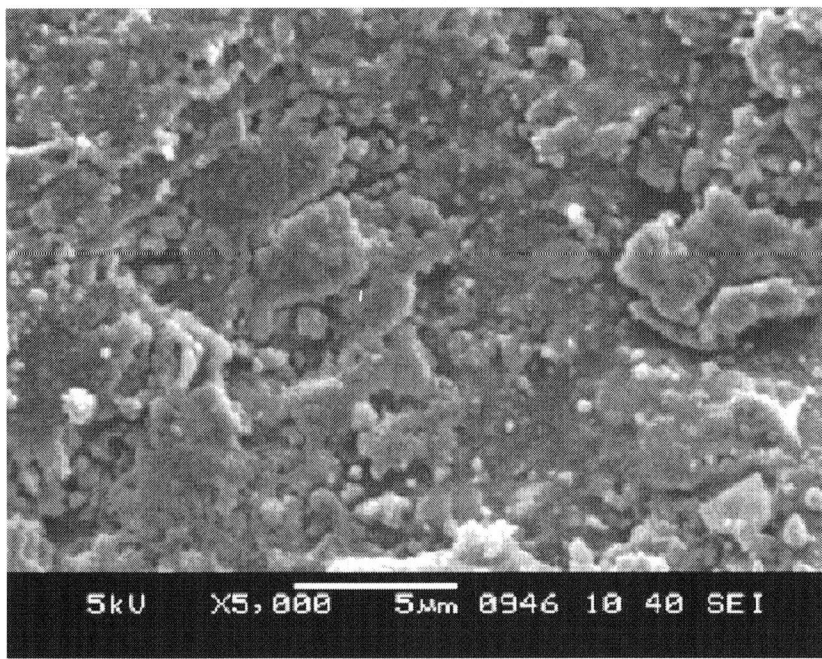

(b)

**Figure 5**: SEM micrographs of huntite and hydromagnesite reinforced plastic composite material having (a) 10 μm, (b) 1 μm, and (c) 0.1 μm particle size.

# Electrical Properties

The most striking property is dielectric behavior of a composite material. The addition of huntite and hydromagnesite powder to tailor dielectric properties is extensively exploited in plastic matrix composite material. From this point of view dielectric constant test result is depicted in Figure 6. It is clear from the figure that the biggest dielectric constant has been achieved with medium size (1 μm). For the loading level test, it has been gotten from the sample with 67% flame retardant powder.

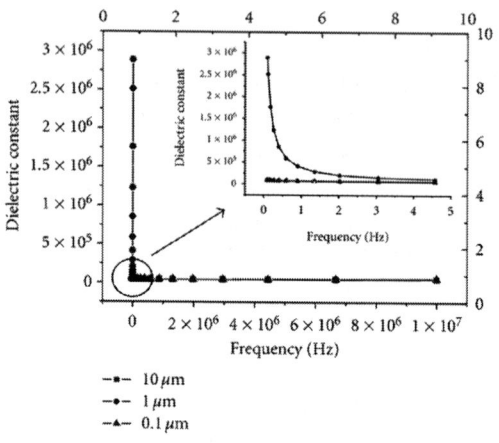

(a)

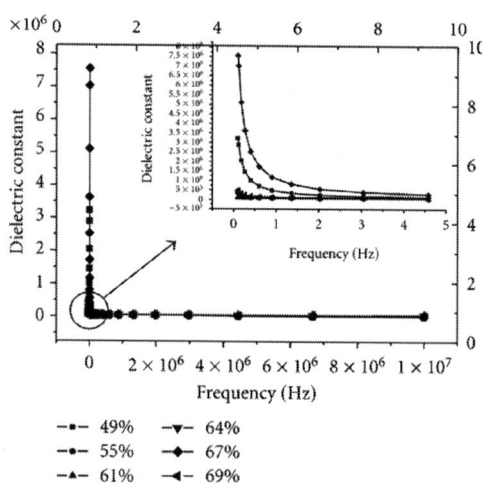

(b)

**Figure 6**: Dielectric constant values of huntite/hydromagnesite reinforced polymeric matrix composite materials as a function of frequency according to (a) particle sizes and (b) contents of reinforced powder.

Actually, the filler particles present in the polymer matrix may be considered as micro/nanocapacitors. The increase in filler loading in the polymer matrix increases the number of such capacitors, which in turn leads to the increase in dielectric constant [29]. At the higher frequency values, just after 10 Hz, dielectric constant became stable at the 40.000 level. On the other hand, for all samples dielectric contents decreased by increasing frequency as similar in [29–32]. At lower frequencies dielectric constants attain high values and then decrease exponentially with increase in frequency. This behavior clearly indicates that the effect of interfacial polarization becomes more and more predominant at lower frequency [29].

Tan (Delta) test result is given in Figure 7. All composites showed a decrease in dissipation factors with increase in frequency. This may be attributed to the dipole relaxation phenomena, where movement of the electric dipoles was not possible to be in phase with the frequency of the applied electric field [30]. The decrease in loss factor with increase in frequency can also be explained by the fact that as the frequency is raised, the dipoles get very less time to orient themselves in the direction of the alternating field [29]. The orientation of the dipoles takes place by the compensation of some electrical energy, which accounts for dielectric loss. At lower frequencies, the effect of orientation polarization of dipoles is higher for getting sufficient time to orient them, thus accounting for higher dielectric loss at lower frequencies [32]. The sample consisting of medium size of mineral showed the biggest value of dissipation factor. For the loading level, the highest dissipation factor was obtained with 49% and 67% loading levels in between 0–5 Hz. Dissipation factor exhibited high values and formed in the low-frequency range. When the frequency was increased; it decreased quickly. As the filler content is increased, heterogeneity of the system is increased, this produces extended interfaces, and results in increased conductivity and higher losses [31].

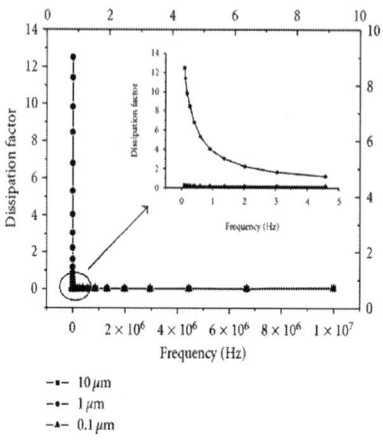

(a)

(b)

**Figure 7**: Dissipation factor of huntite/hydromagnesite reinforced polymeric composite materials as a function of frequency according to (a) particle sizes and (b) contents of reinforced powder.

Figure 8 denotes specific resistance of huntite/hydromagnesite reinforced polymer composites as a function of frequency. The biggest resistivity was obtained with the loading level of 49%. In general, the conductivity of filled polymer composites is governed by both the mechanism of conduction theory (formation of continuous conductive networks) and hopping mechanism (electric field radiation) of conduction theory [24, 25, 27]. Before percolation the conductivity in polymer composite is due to the hopping (jumping) of electrons from one conducting site to another, which is facilitated as the distance between the conducting sites is reduced [32].On the other hand, as explained in [33], the electrical properties of composites were directly related to the morphology of conductive networks, that is, the localization of conductive fillers in the composites.

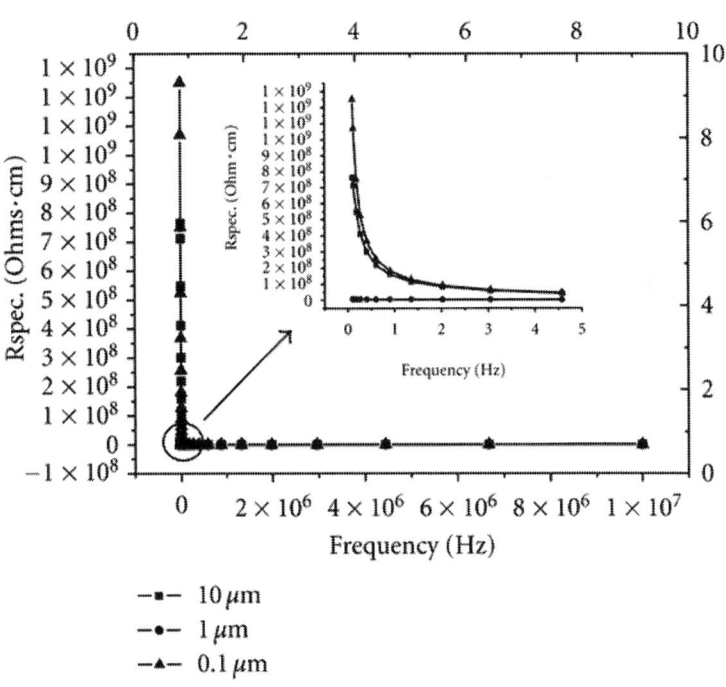

(a)

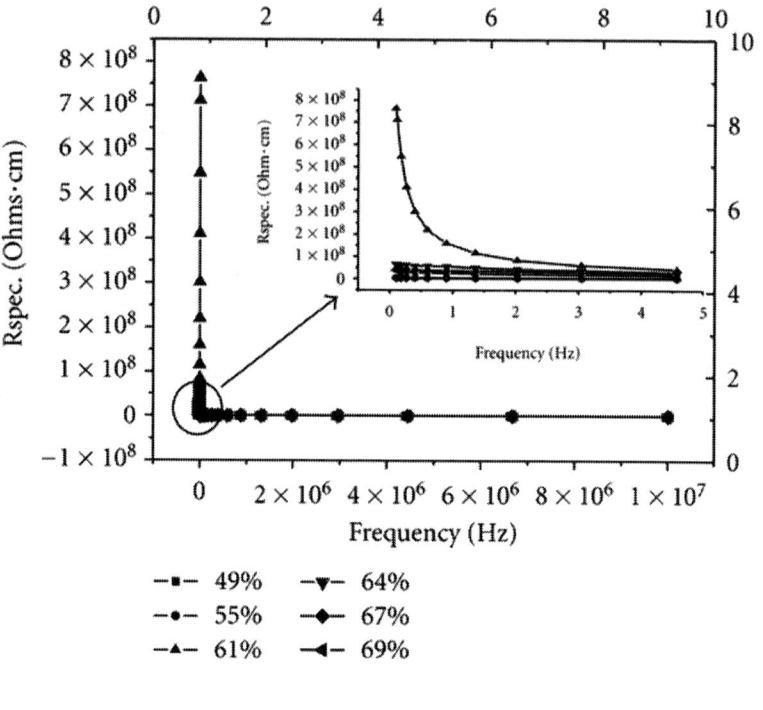

(b)

**Figure 8**: Specific resistances of huntite/hydromagnesite reinforced polymeric composite materials as a function of frequency according to (a) particle sizes and (b) contents of reinforced powder.

Conductivity test result of the flame retardant composite materials is shown in Figure 9. Generally speaking, it should be noted that the complex conductivity increased as a function of frequency as the same in [29, 30]. If the particle size dependence of electrical conductivity of huntite and hydromagnesite reinforced plastic composite materials is to be accounted for, it is necessary for analyses how conductivity depends upon the particle size of reinforced material from the micron scale to nanoscale. The smallest size, which is the highest surface area, is expected to be much better particle-particle the highest conductivity at the percolation threshold [29]. As expected, the resulting characteristic depicted that decreasing the size to nanoscale makes the polymer composite more conductive. One of the notable features of the composite material is the amount of reinforced material. In spite of the fact that it seems to be changing the conductivity related with the loading level, it can be expressed that increasing filler amount increased the polymer's conductivity. The increase in conductivity with the increasing of the filler amount mainly stems from the establishing of conducting networks in the polymer matrix [29, 30, 33]. Besides, we have a good agreement with the literature [34] that finer particles may support this mechanism as the ionic conductivity of the polymer composite increased. In other words, for both size and the loading level effect tests, it can be seen that frequency assists to increase conductivity of the composites.

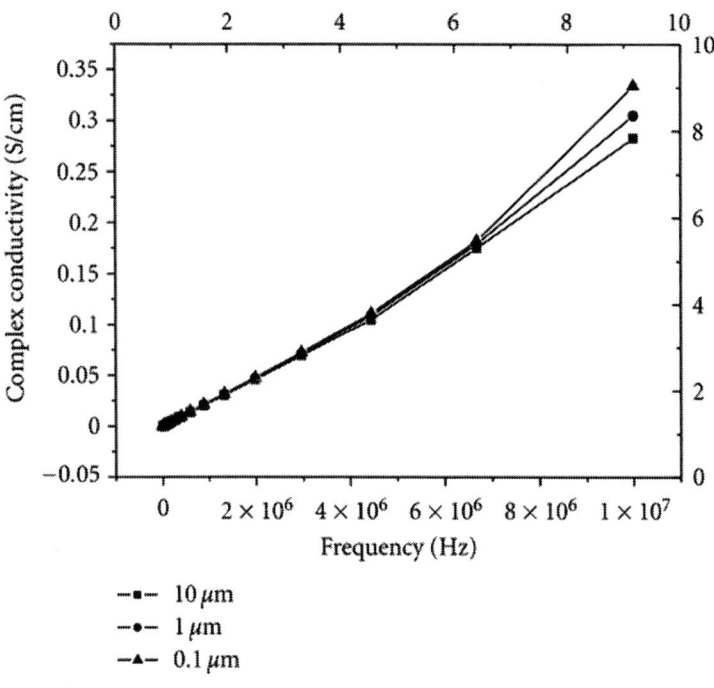

(a)

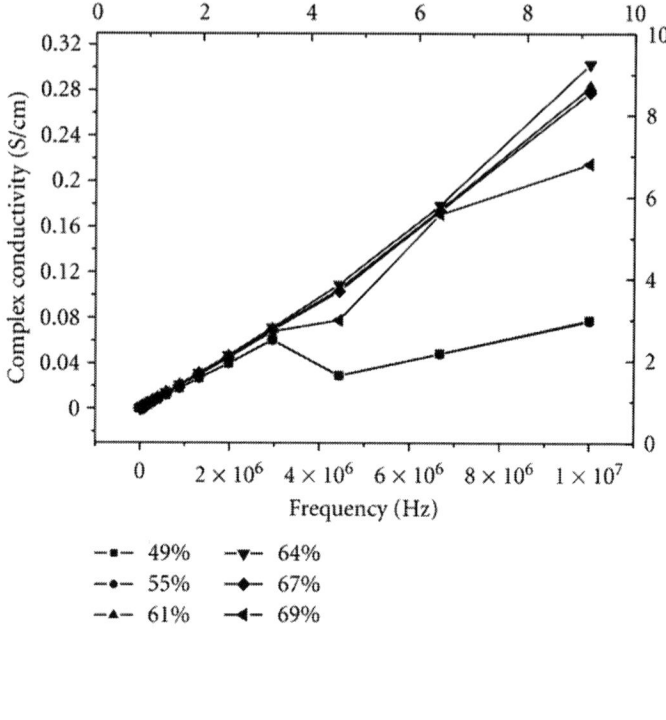

(b)

**Figure 9**: Conductivity of huntite/hydromagnesite reinforced plastic composite materials as a function of frequency according to (a) particle sizes and (b) contents of reinforced powder.

On the other hand, when very low frequency was applied to the composites, the conductivity results were different as seen from Figure 9. In between 0–5 Hz frequency, medium particle content composites have the biggest conductivity. Actually these results are quite similar to the dielectric constant, specific resistance, and tan delta results.

Besides different sizes and different loading level measurements, conductivity tests were carried out for Ag-coated polymer to see if there are any pores or displacements in the polymer composite structure. The obtained results are shown in Figure 10. According to these figures, it can be said that polymer composites had certain amount of pores, as conductivity increased by Ag coating.

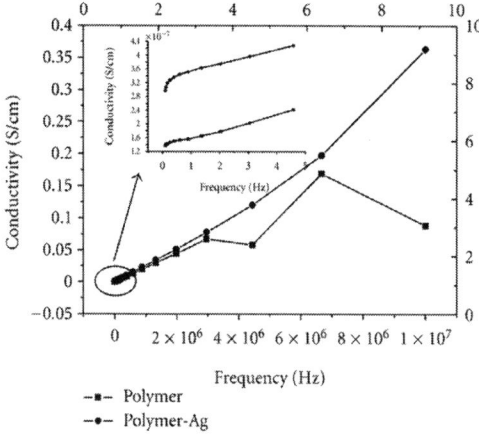

**Figure 10**: Conductivity of Ag-coated and uncoated polymeric composite materials.

# CONCLUSIONS

As huntite and hydromagnesite mineral showed an effective flame retardant behavior with an increase of the time to ignition in the previous study, it was also indicated that decreasing the size increases the flame retardancy property. In this context it can be said that potential escape time will increase by using this kind of polymeric material in a dangerous fire. It is in continuation of the study in this work that the electrical properties were investigated and following results were found; except for the conductivity, dielectric constant, specific resistance, and dissipation factors were decreased by increasing frequency. But conductivity increased with frequency. In addition conductivity increased with decreasing particle amount. With increasing mineral content, dielectric constant and specific resistance values increased. It was shown in the particle size tests that, apart from conductivity, coarser and finer sizes (10 μm and 0.1 μm) behave similarly unlike medium size (1 μm). This may be explained with non-homogenous blending or agglomeration in the 0.1 μm size material. On the other hand, it was seen from the Ag-coated samples tests that composites had some pores that Ag-coated samples showed different electrical property against uncoated ones.

# ACKNOWLEDGMENTS

The authors would like to acknowledge Likya Minelco Madencilik Sti. and Minelco Specialities Limited. They specially thank M. F. Ebeo Iugil and I. Birlik for helping characterization systems.

# REFERENCES

1. N. G. McCrum, C. P. Buckley, and C. B. Bucknall, Principles of Polymer Engineering, Oxford University Press, Oxford, UK, 1997.

2. P. C. Painter and M. M. Coleman, Essentials of Polymer Science and Engineering, 2008.

3. "Introduction to polymers," Copolymers, 2010, http://openlearn.open.ac.uk/mod/oucontent/view.php?id=397829&section=2.3.6.

4. M. O'Driscoll, "Plastic compounding, Where mineral meets polymer," Industrial Minerals, December 1994.

5. F. Laoutid, L. Bonnaud, M. Alexandre, J. M. Lopez-Cuesta, and P. Dubois, "New prospects in flame retardant polymer materials: from fundamentals to nanocomposites," Materials Science and Engineering R, vol. 63, no. 3, pp. 100–125, 2009.

6. A. F. Grand and C. A. Wilkie, Fire Retardancy of Polymeric Materials, New York, NY, USA, 2000.

7. A. A. Basfar and H. J. Bae, "Influence of magnesium hydroxide and huntite hydromagnesite on mechanical properties of ethylene vinyl acetate compounds crosslinked by dicumyl peroxide and ionizing radiation," Journal of Fire Sciences, vol. 28, no. 2, pp. 161–180, 2010.

8. L. Haurie, A. I. Fernández, J. I. Velasco, J. M. Chimenos, J. M. L. Cuesta, and F. Espiell, "Synthetic hydromagnesite as flame retardant. Evaluation of the flame behaviour in a polyethylene matrix,"Polymer Degradation and Stability, vol. 91, no. 5, pp. 989–994, 2006.

9. M. L. Bras, S. Bourbigot, S. Duquesne, C. Jama, and C. Wilkie, Fire Retardancy of Polymers New Applications of Mineral Fillers, 2005.

10.  A. B. Morgan, J. M. Cogen, R. S. Opperman, and J. D. Harris, "The effectiveness of magnesium carbonate-based flame retardants for poly (ethylene-co-vinyl acetate) and poly (ethylene-co-ethyl acrylate)," Fire and Materials, vol. 31, no. 6, pp. 387–410, 2007.

11.  R. Scidt, "In the line of fire, flame retardants overview," Industrial Minerals, pp. 37–41, February 1999.

12.  M. Weber, "Mineral flame retardants, overview & future trends," in Proceedings of the European Minerals & Markets (Euromin '99), pp. 8–10, Nice, France, June 1999.

13.  R. J. Mureinik, "Flame retardants, minerals' growth in plastics," in Proceedings of the Industrial Minerals Information Ltd., IMIL Conference (Euromin '97), pp. 8–10, Barcelona, Spain, June 1997.

14.  M. Xanthos, Functional Fillers for Plastic, Wiley, New York, NY, USA, 2004.

15.  A. I. Fernández, L. Haurie, J. Formosa, J. M. Chimenos, M. Antunes, and J. I. Velasco, "Characterization of poly(ethylene-co-vinyl acetate) (EVA) filled with low grade magnesium hydroxide," Polymer Degradation and Stability, vol. 94, no. 1, pp. 57–60, 2009.

16.  J. H. Schut, "Nanocomposites Do More with Less," Plastics Technology-http://www.PTOOnline.com, 2009.

17.  H. Y. Atay and E. Çelik, "Use of Turkish huntite/hydromagnesite mineral in plastic materials as a flame retardant," Polymer Composites, vol. 31, no. 10, pp. 1692–1700, 2010.

18.  R. N. Rothon, "General principles guiding selection and use of particulate materials," in Particulate-Filled Polymer Composites, Rapra, Shropshire-UK, 2nd edition, 2003.

19.  W. D. Callister, Materials Science and Engineering, John Wiley & Sons, New York, NY, USA, 6th edition, 2003.

20.  A. R. Blythe, Electrical Properties of Polymers, Cambridge Solid State Science Series, 1979.

21.  http://www.zeusinc.com/UserFiles/zeusinc/Documents/Zeus_Dielectric.pdf.

22.  P. Horowitz and W. Hill, The Art of Electronics, Cambridge University Press, 1989.

23. A. Kennelly, "Impedance," American Institute of Electrical Engineers (AIEE), 1893.

24. S. Musikant, What Every Engineer Should Know about Ceramics, CRC Press, 1991.

25. S. Ramo, J. R. Whinnery, and T. V. Duzer, Fields and Waves in Communication Electronics, John Wiley & Sons, New York, NY, USA, 3rd edition, 1994.

26. M. Z. Camur and H. Mutlu, "Major-ion geochemistry and mineralogy of the Salt Lake (Tuz Gölü) basin, Turkey," Chemical Geology, vol. 127, no. 4, pp. 313–329, 1996.

27. W. M. Last, "Petrology of modern carbonate hardgrounds from East Basin lake, a saline maar lake, Southern Australia," Sedimentary Geology, vol. 81, no. 3-4, pp. 215–229, 1992.

28. L. Haurie, A. I. Fernandez, J. I. Velasco, J. M. Chimenos, J. M. Lopez-Cuesta, and F. Espiell, "Effects of milling on the thermal stability of synthetic hydromagnesite," Materials Research Bulletin, vol. 42, no. 6, pp. 1010–1018, 2007.

29. N. J. S. Sohi, M. Rahaman, and D. Khastgir, "Dielectric property and electromagnetic interference shielding effectiveness of ethylene vinyl acetate-based conductive composites: effect of different type of carbon fillers," Polymer Composites, vol. 32, no. 7, pp. 1148–1154, 2011.

30. N. K. Shrivastava and B. B. Khatua, "Development of electrical conductivity with minimum possible percolation threshold in multi-wall carbon nanotube/polystyrene composites," Carbon, vol. 49, no. 13, pp. 4571–4579, 2011.

31. B. K. Singh, P. Kar, Nilesh K. Shrivastava, and B. B. Khatua, "Electrical and mechanical properties of ABS/MWCNT nanocomposites prepared by melt-blending," Journal of Applied Polymer Science, vol. 124, no. 4, pp. 3165–3174, 2012.

32. S. Thomas, P. Abdullateef, A. A. Al-Harthi, et al., "Electrical properties of natural rubber nanocomposites: effect of 1-octadecanol functionalization of carbon nanotubes," Journal of Materials Science, vol. 47, no. 7, pp. 3344–3349, 2012.

33. G. Chen, J. Lu, and D. Wu, "The electrical properties of graphite nanosheet filled immiscible polymer blends," Materials Chemistry and Physics, vol. 104, no. 2-3, pp. 240–243, 2007.

34.  Z. Wen, T. Itoh, T. Uno, M. Kubo, and O. Yamamoto, "Thermal, electrical, and mechanical properties of composite polymer electrolytes based on cross-linked poly(ethylene oxide-co-propylene oxide) and ceramic filler," Solid State Ionics, vol. 160, no. 1-2, pp. 141–148, 2003.

# Dechlorination of Trichloroethylene in Groundwater by Nanoscale Bimetallic Fe/Pd Particles

Tielong LI[1,2], Shujing LI[1], Yongchao LI[1], and Zhaohui JIN[1]

[1]College of Environmental Science and Engineering, Nankai University, Tianjin, China

[2]Tianjin Key Laboratory of Environmental Remediation and Pollution Control/Ministry of Education Key Laboratory of Environmental Processes and Criteria, Nankai University, Tianjin, China

## ABSTRACT

Palladium/iron bimetallic nanoparticles were synthesized using microemulsion method in the water-in-oil (W/O) microemulsion system, which was made up of iso-octane, cetyltrimethyl-ammonium bromide (CTAB), butanol and water and characterized by measuring the conductivity of the solution. Transmission electron microscope (TEM) and energy dispersive X-ray microanalysis (EDX) analysis showed that the average diameter of synthesized palladium/iron bimetallic nanoparticles was less than 80 nm, which was much smaller than the particles produced by the solution method. The palladium/iron bimetallic nanoscale particles produced in the laboratory showed better performance on dechlorinating TCE than the other materials. The nanoscale Fe/Pd particles exhibited high reactivity. When Pd content is 0.5%, the best TCE dechlorination efficiency is achieved within 30min. And Fe/Pd nanoparticles show persistent reaction activity in some sense.

# INTRODUCTION

Chlorinated hydrocarbons are dense non-aqueous phase liquids (DNAPLs), which are utilized by a number of industries as solvents in large quantities on a regular basis. Given this high frequency of use, handling, and transportation, along with past disposal and storage practices, DNAPL compounds presently represent a significant threat to soil and groundwater resources. Trichloroethylene (TCE), has been widely detected in areas adjacent to dry cleaners, automobile manufacturers or shops, asphalt processing plants, and military bases, listed as a priority pollutant by the US EPA, is one of the most commonly detected chlorinated organic compounds in surface water, groundwater and soil. Cleanup of soils and groundwater contaminated by chlorinated hydrocarbons such as TCE and PCBs has been a challenging task for decades.

Remediation of DNAPL-contaminated sites is especially critical because DNAPLs in the subsurface represent a long-term source for groundwater contamination. Aquifers contaminated with DNAPLs are extremely difficult to remediate with standard pump-and-treat methods. Since DNAPLs are extremely difficult and costly to remove, a cost-effective, reliable technology is needed to treat DNAPL contaminants.

Various technologies have been explored for dechlorination of TCE, including bioremediation [1], thermal treatment [2], and permeable reactive barriers [3]. Among many technologies tested so far, abiotic dechlorination using zerovalent iron, $Fe^0$, particles appears to be one of the most promising technologies [4, 5]. The most common metal being utilized for this purpose is iron due to its dehalogenation efficiency, cost, and benign environmental impact. However, due to the limited reactivity, the TCE reduction rate of granular iron particles has been found very slow, with half-lives in the order of days or longer [6,7]. As a result, toxic intermediate byproducts such as vinyl chloride (VC) are often detected [8, 9]. In order to further enhance the dechlorination reaction rates and minimize byproduct formation, bi-metal systems of palladized iron (Fe complex) were developed and used [10–12]. Various strategies have been explored to enhance the dechlorination rates using $Fe^0$-based particles. Because dechlorination of chlorinated compounds by $Fe^0$-based particles is a surface-mediated process, increasing the surface area of iron will increase the dechlorination rate. Therefore, reducing particle size can greatly enhance the degradation rate. Coating iron particles with a second catalytic metal such as Pd, Pt, Ag, or Ni can also accelerate the dechlorination process and thereby prevent formation of toxic byproducts [13–15]. Zhang et al. reported that reducing Pd-coated iron particle size from millimeters to nanometers (10-100 nm) increased TCE degradation rate by 10-100 times. Typically, $Fe^0$-based nanoparticles were prepared by reducing Fe (II) or Fe (III) in aqueous solution using a strong reducing agent (e.g., sodium borohydride, $NaBH_4$). However, due to the extremely high reactivity, the initially formed nanoparticles tend to either react rapidly with surrounding media (e.g., dissolved oxygen (DO) or water) or agglomerate rapidly, resulting in the formation of much larger (in the micrometer to millimeter scale) particles or flocs and rapid loss in reactivity [16]. To overcome these drawbacks, a microemulsion [17, 18] with the cetyltrimethyl-ammonium bromide (CTAB) as the surfactant was used in the preparation of the Fe/Pd nanoparticles and the chemical reactivity for degradation of TCE in water was studied in the experiment of dechlorination in this study. And the nanoparticles were characterized by using the transmission electron microscope (TEM) and energy dispersive X-ray microanalysis (EDX). The water-in-oil (W/O) microemulsion system was made up of iso-octane, CTAB, butanol and water, and was characterized by measuring the conductivity of the solution. The average diameter of synthesized palladium/iron bimetallic nanoparticles using the microemulsion method was less than 80 nm, which was much smaller than the particles produced by the solution method. And then the primary objective of this work was to 1) prepare the nanoparticles in the CTAB microemulsion under inert conditions; and 2) characterize the resultant nanoparticles with respect to their physical stability and chemical reactivity for degradation of TCE in water. And the performance of the nanoparticles produced in the laboratory was studied by dechlorinating TCE. The degradation rate of TCE by the nano palladium/iron bimetallic particles produced in the laboratory was quantified and compared with nano iron particles.

# EXPERIMENTS

The experiments were divided into five parts. And laboratory grade chemicals were used in this experimental study.

## Conductivity

The conductivity was measured using a microprocessor conductometer (DDS-307, ShangHai Precision and Scientific Instrument Corporation). The conductivity was measured to an accuracy of ±0.5% of full scale reading within each range.

## Titration

In order to construct the phase diagram of water/CTAB +n-butanol/ iso-octane system, the titration method was used. Based on the literature review and preliminary tests, the mass ratio of co-surfactant-to-surfactant was fixed, and a predetermined amount of hydrocarbon iso-octane was added to the surfactant mixture. Water was then titrated into the octane–surfactant mixture (octane/ (octane+surfactant) weight ratio of (S)) and changes were observed by visual inspection and conductivity measurement and the information was used in developing a phase diagram.

## Production of Fe/Pd Nanoparticles

The preparation of iron nanoparticles was achieved by mixing rapidly the same volumes of two W/O microemulsion solutions, with $FeSO_4$ solubilized in one solution and $NaBH_4$ as the reducing agent in the other solution. Then firstly, two water/CTAB/isooctane microemulsions

(A and B), differing only in the type of aqueous phase, were prepared. The CTAB was selected as surfactant, with n-butanol as the co-surfactant and isooctane as the oil phase. A solution of 0.2 M $FeSO_4$ was added to the above mixture. The compositions of the microemulsion system used are summarized in Table 1. The weight ratio of isooctane to surfactant was 3. After completely stirring, a transparent yellow solution microemulsion A was obtained. With the same method microemulsion B was prepared, whereas the aqueous phase of microemulsion B was 1.3mol/L $KBH_4$ solution.

And then microemulsions A and B were quickly mixed at the protection of argon in a conical flask for preparation of nanoiron. The reduction reaction made the solution turbid with gas production and with a black color solid dispersed in the solution. The reduction reaction could be expressed as

$$Fe^{2+} + 2BH_4^- + 6H_2O \rightarrow Fe + 2B(OH)_3 + 7H_2\uparrow \quad (1)$$

After the gas evolution, the mixtures were stirred for another 20 min with vigorous stirring (3000r/min). And then loading of Pd to the $Fe^0$ particles was accomplished by adding known quantities of $PdCl_2$ into the microemulsion-$Fe^0$ solution and allowing for reaction for 30 min under vacuum. The amount of Pd added in this study

**Table 1:** Compositions of microemulsions system

|  | Surfactant | Co-Surfactant | Oil | Aqueous |
|---|---|---|---|---|
| ME | CTAB | butanol | iso-octane | $FeSO_4$ |
| Weight (%) | 11 | 10 | 31 | 48 |

was 0.5% (w/w) of Fe. Palladium was deposited on the iron surface through the following redox reaction

$$Pd^{2+} + 2Fe \rightarrow Pd + Fe^{2+} \quad (2)$$

At the end of the reaction, the microemulsion system with the well-dispersed iron nanoparticles was transferred into a sealed vessel filled with argon by a double-tipped needle. The resulting black-gray solids were settled by magnet, and the supernatant was decanted. Then the solids were washed with deionized water and finally with a mixture of anhydrous ethanol and acetone (volume ratio is 1:1) for six times, respectively. Finally, the resultant black-gray solids were dried under argon atomosphere, and then stored in another sealed vessel filled with argon.

# Characterization of Particles

The particles obtained using the microemulsion were characterized using the Philips EM400ST transmission electron microscope (TEM) and energy dispersive X-ray microanalysis (EDX).

# Dechlorination of TCE

Pd content can affect TCE dechlorination efficiency strongly. In order to select the optimal Fe/Pd radio, experiments were designed to investigate the effect of Pd content on TCE dechlorination efficiency of nanoscale Pd/Fe within 30min. In this study, Pd/Fe ratios were 0.2%, 0.5%, 1%, 2% respectively. For comparison, Fe and Fe/Pd nanoparticles were also prepared following similar procedures but use the liquid reduction method. Experiments were conducted to investigate the reduction of TCE using the two synthesized iron particles. Dechlorination of 15 mg/L TCE solution with the 1.5 g/L iron-to-solution loading was investigated. TCE solutions (173 mL) were mixed with the iron particles in separate bottles with Teflon coated caps. The mix was stirred using a platform shaker operated at 200 rpm. The concentration of TCE was analyzed using a HP 6890 GC equipped with an electron capture detector (ECD).

Experiments were also designed to evaluate Fe/Pd nanoparticles' persistent activity. Batch experiments were conducted in 175 ml serum bottles. In each batch bottle, 45μL TCE stock solution was repeatedly spiked into 173ml deionized water which was deoxygenized with Argon. Initial TCE concentration is 56g/L. The solution contained 0.2768g of Fe/Pd nanoparticles.

# RESULTS AND DISCUSSION

## Phase Behavior of Water/Mixed-Surfactants/ Isooctane Microemulsion System

In order to construct the phase diagram of water/mixedsurfactants/ iso-octane microemulsion system, the titration method was used. The optimum weight ratio mentioned above was used to prepare the mixed-surfactants/ n-butanol mixture (EM). A predetermined amount of hydrocarbon iso-octane was added to the surfactant mixture. Water was then titrated into the iso-octane/surfactant mixture and changes were observed by visual inspection and conductivity measurement and the information was used to develop a phase diagram. The phase behavior of the water/mixed-surfactants/iso-octane system is represented in Figure 1. As can be seen in Figure 1, an ideal W/O region has been observed where n-butanol: mixedsurfactants=1:2 (weight ratio), which will be used as the microemulsion system to produce nanoscale particles.

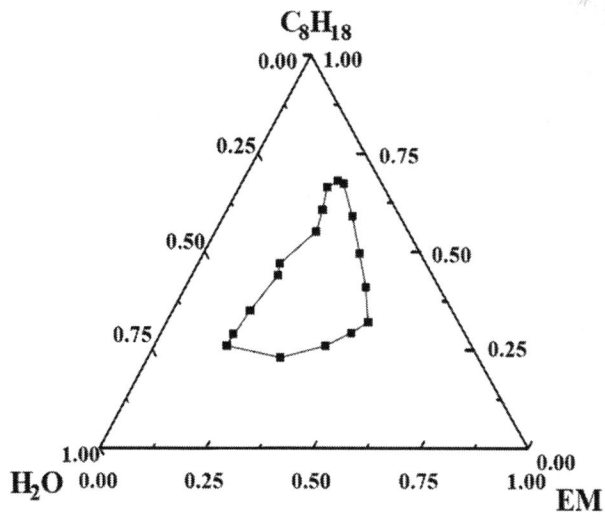

**Figure 1:** Microemulsion phase diagram of CTAB/Butanol/ Octane/H$_2$O system.

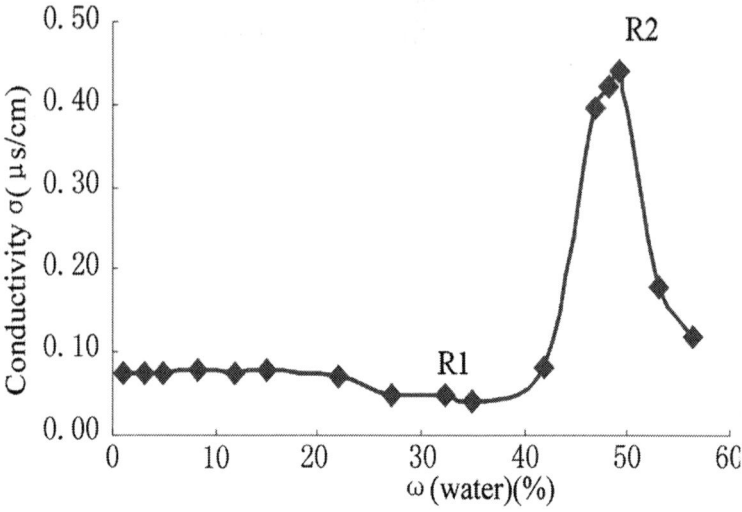

**Figure 2:** Variation of conductivity with amount of water in microemulsion.

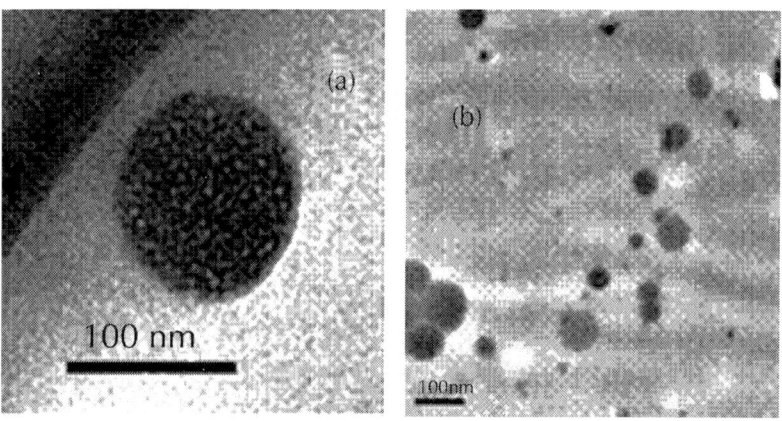

**Figure 3:** Characterizing the nanoscale Fe/Pd particles produced by micro-emulsion method; (a), (b) transmission electron micrograph (TEM) of the particles.

# Conductivity

The variation of conductivity with different water content (W0) is shown in Figure 2. Two turning points (R1, R2) are found in Figure 2. Almost no change was observed in the initial conductivity of the iso-octane/surfactant mixture before R1. This is attributed to the formation of surfactant-bonded water in the system. But after R1 the conductivity increased sharply upon addition of the water into the iso-octane/surfactant mixture since it formed water-in-oil (W/O) microemulsion. In this region, water was present in micelle form and the water droplet size and number increased with an increased amount of water. When the droplet size increased, it was easy to exchange charges between the droplets and hence the conductivity increased. The conductivity reached a maximum when the water content was 50 % (wt%, R2). This can be explained by the saturation of droplets and the percolation of charges through the droplet clusters with minimum resistance. When more water was added to the system after R2, it leads to phase separation and the conductivity was lowered. The optically clear W/O microemulsion system which was formed between R1 and R2 can be used as microreactors for preparation of nanoscale materials.

# Characterization of Synthesized Nanoscale Particles

A small amount of microemulsion containing the palladium/iron bimetallic particles were withdrawn from the reaction vessel, and then preserved in a small sealed glass tube. For the TEM and EDX investigations, a small drop of microemulsion containing iron nanoparticles was deposited on a copper grid.

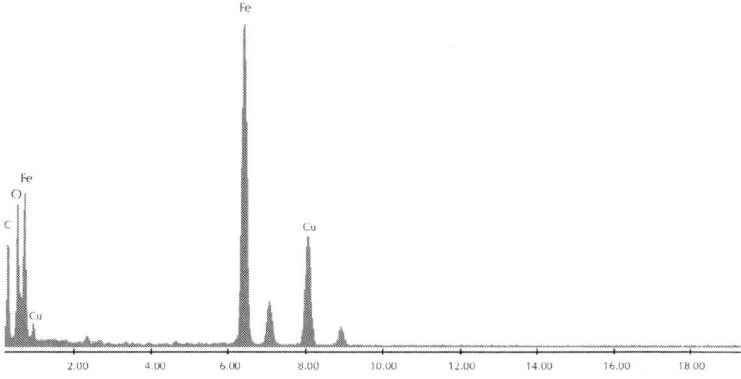

**Figure 4:** Energy dispersive X-ray microanalysis (EDX) of products.

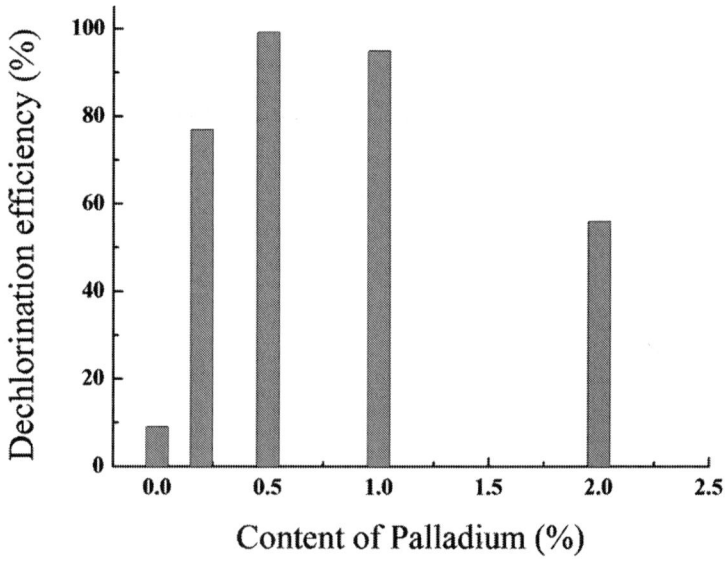

**Figure 5:** Effect of Pd content on TCE dechlorination efficiency of nanoscale Pd/Fe (Reaction conditions: catalyst added 1.6 g/L, T=25°C, pH= 6.98, TCE concentration=14.8 mg/L, stirring at 250rpm, reaction for 30min).

Morphology of the particles is shown in Figure 3(a), (b). From the TEM images, the particles were nearly spherical in shape and uniform in size with the average particle size less than 80 nm. The EDX for the sample is shown in Figure 4. It also shows that the product consisted of Fe and Pd, and no other notable peaks were observed.

# Dechlorination of TCE

As shown in Figure 5, it is apparent that Pd content affects TCE dechlorination efficiency of nanoscale Pd/Fe. By comparison, Fe nanoparticles with Pd showed significant dechlorination. When Pd content was 0.5%, the best TCE dechlorination efficiency was achieved. But, Lower TCE was observed along with Pd content increasing. When Pd content was 2%, TCE dechlorination efficiency was merely 60%. Dechlorination efficiency decreasing was probably attributable to Pd films coating on the surface of Fe nanoparticles closely, that prevented Fe oxidation.

Figure 6 shows that degradation of TCE can be greatly enhanced when a small fraction (0.5% of Fe) of Pd was coated on the Fe particles. At the rather modest Fe dose of 1.5 g/L, even the NPI particles prepared by liquid method were able to eliminate 61% of TCE in the batch reactor within 0.5 h. When the particles were prepared by microemulsion, 98% of TCE was destroyed within 0.5 h. In contrast, a TCE degradation of only 8% was observed when the nanoscale Fe particles were prepared by microemulsion.

It can conclude that dechlorination efficiency of trichloroethylene with Fe/Pd nanoparticles is above 99% in 30min from Figure 7. It is also expected that the nanoparticles can remain reactive for extended period of time in the subsurface environment.

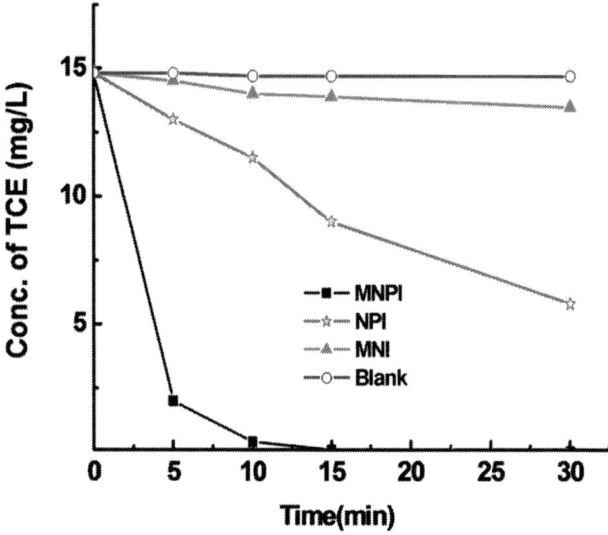

**Figure 6:** Dechlorination of TCE using nanoscale iron particles (MNI), nanoscale Fe/Pd particles (MNPI) prepared in microemulsion method, and nanoscale Fe/Pd particles prepared in liquid method (NPI). Fe dose was 1.5 g/L in all cases. Pd-to-iron ratio was 0.5/100 (w/w).

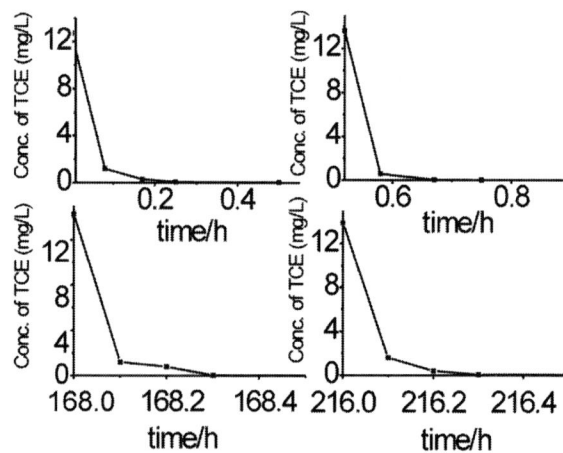

**Figure 7:** TCE concentration change with time in a multiple spiking experiment (reaction conditions: Fe/Pd added 1.6 g/L, T=25°C, pH=6.98, TCE concentration=14.8 mg/L, stirring at 250rpm).

# CONCLUSIONS

Based on the experimental study, the following conclusions are advanced:

Iron and Fe/Pd nanoparticles were successfully produced by the solution method and the microemulsion method. For the microemulsion method, the particles were uniform in size and were less than 80 nm, and appear to be clearly discrete and well-dispersed.

It showed in our results that the Fe/Pd nanoparticles synthesized in our lab have the potential of remediation of TCE. The reduction rate of TCE by the microemulsion Fe/Pd product was the highest compared with the solution product and to the iron prepared by microemulsion. Particle size is an important factor in determining the degradation rate of TCE..

The nanoscale Fe/Pd particles exhibited high reactivity. When Pd content is 0.5%, the best TCE dechlorination efficiency is achieved. And Fe/Pd nanoparticles show persistent reaction activity in some sense.

# ACKNOWLEDGMENTS

This work was supported by Doctoral Fund for Young Scholar of Ministry of Education of China (No. 20070055053).

# REFERENCES

1.    J. MunakataMarr, P. L. McCarty, and M. S. Shields, "Enhancement of trichloroethylene degradation in aquifer microcosms bioaugmented with wild type and genetically altered Burkholderia (Pseudomonas) cepacia G4 and PR1," Environmental Science and Technology, Vol. 30, No. 6, pp. 2045–2052, 1996.

2.    A. K. Friis, H. J. Albrechtsen, and G. Heron, "Redox processes and release of organic matter after thermal treatment of a TCE-Contaminated aquifer," Environmental Science and Technology, Vol. 39, No. 15, pp. 5787–5795, 2005.

3.   H. Shen and J. T. Wilson, "Trichloroethylene removal from groundwater in flow-through columns simulating a permeable reactive barrier constructed with plant mulch," Environmental Science and Technology, Vol. 41, No. 11, pp. 4077–4083, 2007.

4.   L. J. Matheson and P. G. Tratnyek, "Reductive dehalogenation of chlorinated methanes by iron metal," Environmental Science and Technology, Vol. 28, No. 12, pp. 2045–2053, 1994.

5.   C. B. Wang and W. X. Zhang, "Synthesizing nanoscale iron particles for rapid and complete dechlorination of TCE and PCBs." Environmental Science and Technology, Vol. 31, No. 7, pp. 2154–2156, 1997.

6.   T. L. Johnson, M. M. Scherer, and P. G. Tratnyek, "Kinetics of halogenated organic compound degradation by iron metal," Environmental Science and Technology, Vol. 30, No. 8, pp. 2634–2640, 1996.

7.   G. D. Sayles, G. You, M. Wang, and M. J. Kupferle, "DDT, DDD, and DDE dechlorination by zerovalent iron," Environmental Science and Technology, Vol. 31, No. 12, pp. 3448–3454, 1997.

8.   W. A. Arnold and A. L. Roberts, "Pathways and kinetics of chlorinated ethylene and chlorinated acetylene reaction with Fe (0) particles," Environmental Science and Technology, Vol. 34, No. 9, pp. 1794–1805, 2000.

9.   W. S. Orth and R. W. Gillham, "Dechlorination of trichloroethene in aqueous solution using $Fe^0$," Environmental Science and Technology, Vol. 30, No. 1, pp. 66–71, 1996.

10.  T. Li and J. Farrell, "Reductive dechlorination of trichloroethene and carbon tetrachloride using iron and palladized-iron cathodes," Environmental Science and Technology, Vol. 34, No. 1, pp. 173–179, 2000.

11.  P. Zhang, X. Tao, Z. Li, and R. S. Bowman, ,"Enhanced perchloroethylene reduction in column systems using surfactant-modified zeolite/zero-valent iron pellets," Environmental Science and Technology, Vol. 36, No. 16, pp. 3597–3603, 2002.

12.  G. V. Lowry and M. Reinhard, "Pd-catalyzed TCE dechlorination in groundwater: solute effects, biological control, and oxidative catalyst regeneration," Environmental Science and Technology, Vol. 34, No. 15, pp. 3217–3223, 2000.

13. W. X. Zhang, C. B. Wang, and H. L. Lien, "Treatment of chlorinated organic contaminants with nanoscale bimetallic particles," Catalogue Today, Vol. 40, No. 4, pp. 387–395, and 1998.

14. Y. Xu and W. X. Zhang, "Subcolloidal Fe/Ag particles for reductive dehalogenation of chlorinated benzenes," Industrial And Engineering Chemistry Research, Vol. 39, No. 7, pp. 2238–2244, 2000.

15. B. Schrick, J. L. Blough, A. D. Jones, and T. E. Mallouk, "Hydrodechlorination of trichloroethylene to hydrocarbons using bimetallic nickel-iron nanoparticles," Chemistry of Materials, Vol. 14, No. 12, pp. 5140–5147, 2002.

16. F. He and D. Y. Zhao, "Preparation and characterization of a new class of starch-stabilized bimetallic nanoparticles for degradation of chlorinated hydrocarbons in water," Environmental Science and Technology, Vol. 39, No. 9, pp. 3314–3320, 2005.

17. F. Li, C. Vipulanandan, and K. K. Mohanty, "Microemulsion and solution approaches to nanoparticle iron production of degradation of trichloroethylene," Colloids Surface, Vol. 223, No. 1–3, pp. 103–112, 2003.

18. A. J. Zarur and J. Y. Ying, "Reverse microemulsion synthesis of nanostructured complex oxides for catalytic combustion," Nature, Vol. 403, No. 6, pp. 65–68, 2000.

# Chapter 3

# Advances in Application of Natural Clay and its Composites in Removal of Biological, Organic, and Inorganic Contaminants from Drinking Water

Rajani Srinivasan

Blackland Research and Extension Center, AgriLife Research, Texas A&M University System, 720 East Blackland Road, Temple, TX 76502, USA

## ABSTRACT

Natural clays are abundantly available low-cost natural resource which is nontoxic to ecosystem. Over the recent years, research on the modification of clay to increase their adsorbent capacity to remove other contaminants from drinking water other than metals is in progress.

This paper reviews the recent development of natural clays and their modified forms as adsorbing agents for treating drinking water and their sources. This paper describes the versatile nature of natural clay and their ability to adsorb variety of contaminants ranging from inorganic to emerging, which are present in the drinking water. The properties and modification of the natural clay and its significance in removing a specific type of contaminant are described. The adsorbing efficiency of the natural and modified clay in the purification of drinking water, when compared to existing technologies, materials, and methods was found to be significantly higher or comparable.

# INTRODUCTION

Clean drinking water is one of the implicit requisites for a healthy human population. However, the growing industrialization, and extensive use of chemicals for various concerns, has increased the burden of unwanted pollutants of drinking water in developing and developed countries all over the world. The entry of potentially hazardous substances into the ecosystem is increasing day by day. Problems in drinking water quality include presence of excess fluoride, arsenic and natural organic matters, heavy metals, and variety of pathogens are the major causes for various water-borne diseases. Since it is not possible to prevent these chemicals from draining into the drinking water sources, the only way to maintain safer water bodies is to develop efficient purifying technologies. One such beneficial and successful procedure that has been in use is that of purification of water using natural and modified adsorbents. Several natural adsorbents are being used for treatment of contaminated drinking water and its sources. When a comparison is made with other low-cost adsorbents, the clays and their modified composites have been found to be either better or equivalent in contaminant adsorption capacity from water. Clay minerals are generally categorized into following groups: montmorillonite, smectite, kaolinite, illite, and chlorite. Montmorillonite, kaolinite, and illite are widely used because of their high specific surface area, chemical and mechanical stability, a variety of surface and structural properties, and low cost [1–4]. The price of clay is about $0.005–0.46/kg, and the price of montmorillonite is about $0.04–0.12/kg, which is 20 times cheaper than activated carbon [5, 6].

Clays are hydrous aluminosilicates broadly defined as those minerals that make up the colloid fraction of soils, sediments, rocks, and water [7] and may be composed of mixtures of fine grained clay minerals and clay-sized crystals of other minerals such as quartz, carbonate, and metal oxides. Clays play an important role in the environment by acting as a natural scavenger of pollutants by taking up cations and anions either through ion exchange or adsorption or both. Thus, clays invariably contain exchangeable cations and anions held to the surface. The prominent cations and anions found on clay surface are $Ca^{2+}$, $Mg^{2+}$, $H^+$, $K^+$, $NH_4^+$, $N_a^+$, $SO_4^{2-}$, $Cl^-$, $PO_4^{3-}$, and $NO^{3-}$. These ions can be exchanged with other ions relatively easily without affecting the clay mineral structure. Large specific surface area, chemical and mechanical stability, layered structure, high cation exchange capacity (CEC), and so forth have made the clays excellent adsorbent materials [8]. Both Bronsted and Lewis type of acidity in clays [9] have boosted the adsorption capacity of clay minerals to a great extent. The Bronsted acidity arises from $H^+$ ions on the surface formed by dissociation of water molecules of hydrated exchangeable metal cations on the surface. The Bronsted acidity may also arise if there is a net negative charge on the surface due to the substitution of $Si^{4+}$ by $Al^{3+}$ in some of the tetrahedral positions and the resultant charge is balanced by $H_3O^+$ cations. The Lewis acidity arises from exposed trivalent cations, mostly $Al^{3+}$ at the edges, or $Al^{3+}$ arising from rupture of Si-O-Al bonds, or through dehydroxylation of some Bronsted acid sites. The edges and the faces of clay particles can adsorb anions, cations, and nonionic and polar contaminants from natural water. The contaminants accumulate on clay surface leading to their immobilization through the processes of ion exchange, coordination, or ion-dipole interactions. Sometimes the pollutants can be held through H-bonding, van der Waals interactions, or hydrophobic bonding arising from either strong or weak interactions. The strength of the interactions is determined by various structural and other features of the clay mineral. van Olphen [10] has cited several types of active sites in clays, namely, (i) Bronsted acid or proton donor sites, created by interactions of adsorbed or interlayer water molecules, (ii) Lewis acid or electron acceptor sites occurring due to dehydroxylation, (iii) oxidizing sites, due to the presence of some cations in octahedral positions or due to adsorbed oxygen on surfaces, (iv) reducing sites produced due to the presence of some cations, and (v) surface hydroxyl groups, mostly found in the edges, bound to Si, Al, or other octahedral cations. Clays have been good adsorbents because of the existence of several types of active sites on the surface, which include Bronsted and Lewis acid sites and ion exchange sites. The edge hydroxyl groups have been particularly active for various types of interactions. Clays and modified clays have been found particularly useful for adsorption of heavy metals. Clays have received attention as excellent adsorbents of As, Cd, Cr, Co, Cu, Fe, Pb, Mn, Ni, and Zn in their ionic forms from aqueous medium. The adsorption capacities differs from metal to metal and also depend on the type of clay used [11].

Composites can be defined as natural or synthesized materials made from two or more materials with significantly different physical and chemical properties which remain separate and distinct at the microscopic or macroscopic scale within the material. Composites are synthesized to combine the desired properties of the materials in the composite. In nanocomposite, nanoparticles (clay, metal, carbon nanotubes, etc.) act as fillers in a matrix. Nanoparticles are particles of less than 100 nm in diameter that exhibit new or enhanced size-dependent properties compared with larger particles of the same material. Clay composite or nanocomposites are the materials in which major component of the material is clay in combination with other materials like metals, polymer, and so forth. Recently, the development and characterization of nanostructured polymer-clay composites has received special attention because of their advantages in comparison to the traditional polymer composites. Minimal additions of nanoclay enhance mechanical, thermal, and dimensional and barrier performance properties significantly because of the large contact area between polymer and clay on a nanoscale [12–16].

In the last few years, polymer-clay nanocomposites have received a great deal of attention, including studies on developing the composites as sorbents for nonionic and anionic pollutants [17], organic pollutants [18], anionic herbicide [19], and atrazine [20]. Chitosan-montmorillonite composites have been well characterized [21–23], and the adsorption of anionic pollutants by these composites has been investigated [24, 25].

This paper reviews the recent use of natural clay and its composites as an ecofriendly efficient adsorbent for removal of organic, inorganic, and pathogenic contaminants from drinking water and its sources. The major goals of the paper are

- to conduct comprehensive review of the literature to emphasize the importance of using clay and its modified forms as versatile, environmentally friendly adsorbents for contaminant removal from drinking water and its sources,

- to emphasize on the types of modification on the natural clays and the benefits of these modification on removal of major emerging contaminants of the present time,

- to analyze the quantitative efficiency of the individual clay and its composites in removing the various contaminants and study the effects of variable like pH, temperature, and other conditions limiting or enhancing the adsorbent efficiencies of the clay materials.

The author believes that this paper will help in understanding the efficiency of the low-cost clay materials and their abundant availability as an alternative option to otherwise expensive and in few cases toxic treatment technologies being used globally for drinking water treatment.

An extensive literature review by the author resulted in compilation of several papers. These papers reported the use of either natural clay or their modified composites as adsorbents or technologies for removal of contaminants present in the drinking water or affecting the drinking water sources. An attempt is being made to include sufficient information like the type of natural clays used, their modification, their efficiency and variables affecting them from each work so that complete information is available to the readers. The type of analysis being used to get the maximum removal efficiency of each contaminant is also included. The complete review is being summarized in the Table 1.

**Table 1:** Summary of the advances in application of natural clay and its composites in removing different contaminants from drinking water and its sources

| S. No. | Contaminants in drinking water sources | Type of clay and modification | Reference number | Efficiency | Effecting variable | Value of the effecting variable | Specific comments |
|---|---|---|---|---|---|---|---|
| 1 | Metals | | | | | | |
| | Cadmium | Kaolinite and montmorillonite and their modified forms bentonite | \|26, 27\| | | | | Montmorillonite and its modified forms had higher metal adsorbing capacity |
| | Chromium | | \|29\| | | | | |
| | Cobalt | | \|29, 30\| | | | | |
| | Copper | | \|27, 29,31, 32\| | | | | |
| | Iron | | \|33\| | | | | |
| | Lead | | \|34, 35\| | | | | |
| | Manganese | | \|29\| | | | | |
| | Nickel | | \|29, 32,36\| | | | | |
| | Zinc | | \|31, 37\| | | | | |
| | Nickel Copper Cadmium Zinc | Bentonite clay iron oxide composite | \|38\| | | | | |
| | Lead and Cadmium | Beidellite | \|39\| | 83.3–86.9 mg g−1 42–45.6 mg g−1 | | | |
| | Tungsten | Montmorillonite coated with chitosan | \|40\| | 23.9 mg g−1 | pH | 4 | Efficiency decreases with increase in PH |
| | Uranium | Thermally activated bentonite (TAB) | \|41\| | 196 m1g−1 | Temperature pH | 440°C 9 | Efficiency increases with increase in temp and pH |
| | Lead and Zinc | Bentonite | \|42\| | | Adsorbent dose | 5 gL−1 20gL−1 | Efficiency increases with increase in adsorbent dose. |
| | Hexavalent chromium | Montmorillonite supported magnetic nanoparticle | \|43\| | 15.3 mg g−1 | pH | | |

| Metal | Clay | Ref. | Removal | Parameter | Value | Remarks |
|---|---|---|---|---|---|---|
| Cobalt | Kaolinite and montmorillonite | [45] | | | | |
| Copper Nickel Cobalt Manganese | Kaolinite and montmorillonite | [29] | 11.0 mg g−1 | | | |
| Chromium | Chitosan-montmorillonite-Na (organo-nanoclay composite) | [16] | | pH | 3 | |
| Arsenic | Calcined kaolin and bentonite pretreated with $Fe^{2+}$, $Fe^{3+}$, $Al^{3+}$ and $Mn^{2+}$ | [46] | 92–99% 50% | Type of metal pretreatment | $Fe^{2+}$ and $Mn^{2+}$ $Fe^{3+}$ and $Al^{3+}$ | |
| | Montmorillonite, kaolinite and Illite | [47] | 90% | Concentration of sodium chloride | | |
| Cadmium Chromium, Copper Mercury Lead Zinc | Mixed clay (illite, kaolinite, mixed layer minerals and nonclay mineral carbonate fluoroapatite | [48] | 85% 90% 50% 60% 100% 92% | pH | 6 and 9 5<6.8 pH independent <7.67 <7 | |
| Selenium | Chitosan montmorillonite | [52] | 18.4 mg g−1 | | | |
| Arsenate and Arsenite | Ti-pillared montmorillonite | [53] | Greater than 60% | Temperature pH | 25°C–45°C 5 for As(III) and 3 for As(V) | Arsenite and arsenate removal decreased with increase in temperature but opposite trend was seen in the temperature range of 45°C–65°C in case of arsenate. |
| Lead | Sodium montmorillonite clay-carboxy methyl cellulose composite | [54] | 97% | | | |
| Copper | Bentonite polyacrylamide composite | [55] | | pH | 7 | Increasing temperature and decrease in ionic strength favors copper adsorption. |

| Contaminant | Material | Ref | Capacity/Removal | Parameter | Value | Description |
|---|---|---|---|---|---|---|
| Lead Nickel Cadmium Copper | Bentonite-methylene bis-acrylamide | [56] | 1666.67 mgg$^{-1}$ 270.27 mgg$^{-1}$ 416.67 mgg$^{-1}$ 222.2 mgg$^{-1}$ | | | |
| **II Inorganic contaminants** | | | | | | |
| Fluoride | Magnesium incorporated bentonite magnesium-bentonite manganese-bentonite | [57] | 95.45% No significant removal | Desorption | 97% | Decreases the capacity of the desorbed MB from 95% to 75%. |
| | Lanthanum-bentonite | [58] | 68% | pH | 5 | Fluoride removal decreases at alkaline pH |
| | Zirconium loaded bentonite | [59] | — | pH | Less than 6 | Best removal is found below pH 6. |
| Nitrates | Calcium montmorillonite activated by hydrochloric acid | [60] | 22.28% | Stirring time | 68 hours | 13.74% removal was increased to 22.28% when stirring time increased from 0.5 hours to 68 hours |
| **III Organic contaminants** | | | | | | |
| Dichloroacetic acid | Bentonite-based Absorptive ozonation followed by catalytic oxidation by Fe3+ | [61] | 92% | Addition of Fe3+ | 5 mgL$^{-1}$ | Increase of concentration of Fe3+ from 0.5mgL$^{-1}$ to 5mgL$^{-1}$ increased the removal efficiency from 68% to 92% |
| Carbon tetrachloride | Quaternary ammonium salt-modified bentonite | [62] | 70% | | | |
| Emerging contaminants: naproxen, salicylic acid, clofibric acid and carbamazepine | Inorganic-organic-intercalated (IO) bentonites | [63] | 2.69 μmolg$^{-1}$ 5.55 μmolg$^{-1}$ | Addition of different transition metal | varies | Ni < Cu < Co carbamazepine < clofibric acid < naproxen < salicylic acid |

| | Contaminant | Material | Ref | Removal | Parameter | Value | Remarks |
|---|---|---|---|---|---|---|---|
| | Phenol | Bentonite modified with cationic surfactant, acetyl trimethyl ammonium bromide (CTAB) | [64] | 333 mgg−1 | pH | 9 | |
| | Humid acid and O-dichlorobenzene | Combined ozonation and bentonite coagulation | [65] | 95% of HA and 74% of DCB | Iron | 1–5 mgL−1 | increase of iron from 0–10 mgL−1 in the system improved the adsorption efficiency for both HA and DCB |
| | Algae removal | Montmorillonite KSF | [66] | 100% | Dose of clay | 200 mg/L | |
| | Blue green algae (Cyanobacterial microcystis aeruginosa) | Montmorillonite-Cu2+/Fe3+oxides magnetic material | [67] | 92% | Ratio of Clay: Cu2+/Fe3+ | 2:1 | With increase in the ration from 1:1 to 2:1 removal efficiency increased from 48% to 92% |
| | Atrazine | 4-vinylpyridine-co-styrene-montmorillonite | [68] | 90–99% | | | |
| | Atrazine, sulfentrazone, imazaquin and alachlor | Vesicle-clay complex (Di dodecyldimethylammonium bromide-montmorillonite) | [69] | 60% atrazine and 90–100% for others | Presence of all the contaminants together | | Presence of all the contaminants together had a synergistic effect in their removal |
| | Naphthalene and phenolic derivative | Crystal violet tetraphenyl phosphonium-montmorillonite | [70] | ~99% | Organo clay dose | 1.67 gL−1 | |
| | Salicylic acid | Bentonite and kaolin | [71] | — | — | — | |
| | Phenol nitro benzene | Cetyltrimethyl ammonium bromide | [72] | 150 mgg−1 69 mgg−1 | | | |
| | Carbamazepine | Modified smectite clays | [73] | | | | |
| IV | Pathogen | | | | | | |
| | Microcystin-LR | Natural clay minerals consisting of kaolin and montmorillonite | [74] | 81% | | | — |

The review section is divided into following four headings on the basis of the type of contaminant removed by the clay and its composites: (1) heavy metals (2) inorganic contaminants (3) organic contaminant, and (4) pathogens.

# Heavy Metals

Heavy metal contamination in drinking water resources has serious effects on the health of human beings, animals, and plants. Currently, many researchers are working in this field to find an appropriate solution for removing various metals present in the water. Application and efficiency of different type of natural clay and their composites in removing various metals like arsenic, iron, manganese, lead cadmium, uranium, chromium, selenium tungsten, and zinc are reviewed in the following sections. Clays and their modified forms have received wide attention recently for use as adsorbents of metal ions from aqueous medium because of their easy availability and comparatively less cost. [11].Removal of heavy metals by natural clays and their modified forms, kaolinite and montmorillonite in particular, has been reviewed by Bhattacharyya and Sen Gupta [11]. Their review reports the modification of the above mentioned clays by pillaring with various polyoxy cations of $Zr^{4+}$, $Al^{3+}$, $Si^{4+}$, $Ti^{4+}$, $Fe^{3+}$, $Cr^{3+}$ or $Ga^{3+}$, and so forth. The adsorption of toxic metals, namely, As, Cd, Cr, Co, Cu, Fe, Pb, Mn, Ni, Zn, and so forth, has been studied predominantly. They found montmorillonite and its modified forms have much higher metal adsorption capacity compared to that of kaolinite and its modified forms. Their work reports the successful and improved adsorption of metals like Cd [26, 27], Cr [28], CO [29, 30], Cu [27, 29, 31, 32] Fe [33], Pb [34, 35], Mn, Ni [29, 32, 36], and zinc [31, 37] by kaolinite, montmorillonite, and their modified forms. They found that montmorillonite and its modified forms have higher metal adsorbing capacity as compared to their counterparts. Oliveira et al. [38] prepared clay iron oxide composite for adsorption of metal ions $Ni^{2+}$, $Cu^{2+}$, $Cd^{2+}$, and $Zn^{2+}$ from aqueous solution. They compared the metal adsorption capacity of bentonite clay and its magnetic composite. They showed that the presence of iron oxide increased the adsorption capacity of the bentonite. These adsorbents showed the advantage to be easily removed from the medium by a simple magnetic separation procedure after saturation is reached. Beidellite, a low-cost and natural clay mineral, was used as an adsorbent for the removal of lead and cadmium ions from aqueous solutions in batch experiments by Etci et al. [39]. Beidellite used in this study was obtained from the Eastern Black Sea region in Turkey from small-sized deposits formed from the alteration of volcanic rocks. The X-ray diffractometer (XRD) analysis of the clay was performed. It was found to be beidellite clay mineral of the smectite group. Smectites are 2 : 1 phyllosilicates with a total negative layer charge. The formula of beidellite was found to be $N_a 0.5Al_2(Si, Al)_4O_{10}(OH)_2 \cdot 2H_2O$. The kinetics of adsorption process followed the pseudosecond-order reaction. The adsorption capacities ($Q°$) of beidellite for lead and cadmium ions were calculated from the Langmuir isotherm. It was found that adsorption capacity was in the range of 83.3–86.9 mg g$^{-1}$ for lead and 42–45.6 mg g$^{-1}$ for cadmium at different temperatures. Thermodynamic studies showed that the metal uptake reaction by beidellite was endothermic in nature. Natural montmorillonite clay coated with biopolymer chitosan was compared for its efficiency to remove tungsten from simulated drinking water in Neveda, USA, by Gecol et al. [40] Biopolymer-coated clay particles were synthesized. The effects of tungsten concentration in feed water (20–500 ppm) and water pH (4, 5.5 and 6.4) on the zeta potential of adsorbent particles, tungsten removal, and adsorption equilibrium were studied using chitosan-coated clay and natural clay. It was shown that coating clay particles with chitosan shifts the net surface charge of clay from negative to positive and the point of zero charge (PZC) of clay from 2.8 and 5.8. The net surface charge of biosorbent particles decreases with an increase in the tungsten concentration of feed water, because the positively charged sites are consumed by the adsorption of tungsten anions. Chitosan-coated clay was found to be much more effective than natural clay for the removal of tungsten. The tungsten removal efficiencies of both chitosan-coated clay and natural clay decrease with an increase in the pH level and an increase in the tungsten concentration of the feed water. Adsorption equilibrium studies show that tungsten removal is the highest at pH 4. Adsorption of tungsten species on both chitosan-coated clay and natural clay seemed to obey Langmuir isotherm within the range of concentrations and pH investigated. The maximum tungsten adsorption capacities of chitosan, chitosan coated clay and natural clay were found to be 632 mg tungsten per gram of chitosan, 23.9 mg tungsten per gram of chitosan coated clay, and 5.45 mg tungsten per gram of natural clay at pH 4, respectively. The tungsten species adsorption on chitosan coated clay was found to be effected by the ionic attraction between the protonated surface groups on chitosan and the negatively charged tungsten species. But, the tungsten species adsorption on natural clay is governed by the positively charged clay particle edges formed by broken bonds of Al–O and Si–O.

Aytas et al. [41] investigated the effect of pH, contact time, temperature, and initial metal concentration on uranium (U(VI)) adsorption on thermally activated bentonite (TAB). Graphical correlation of various adsorption isotherm models like Freundlich and Dubinin-Radushkevich were carried out for TAB. Various thermodynamic parameters such as Gibb's free energy, entropy, and enthalpy of the on-going adsorption process was calculated. Surface area, FT-IR, and DTA-TG spectra analyses were carried out to determine the adsorptive characteristic of bentonite sample. It was found that the adsorption properties of bentonites change when the samples are calcined at $350°$–$550°C$, that is, at conditions that the layer structure is retained. When bentonite was calcinated at $400°C$, the adsorptive capacity is highest but decreases when above $400°C$. the reason being that the rise of temperature breaks the crystal structure and decreases the specific surface area and adsorbability. The TAB has a maximum sorption at pH 9.0. The initial $U^{6+}$ concentration was varied from 25 to 125 mg $L^{-1}$ to evaluate its effect on adsorption efficiency. The $U^{6+}$ adsorption increased in the initial concentration range from 25 to 100 mg $L^{-1}$ and slightly decreased after 100 mg $L^{-1}$. The effect of contact time was studied using a constant concentration of uranium solution at $30°C$. The sorption of $U^{6+}$ ions has been investigated in the contact time range of 5–180 minutes. It was shown that the variation of $K_d$ and percentage adsorption with shaking time for $U^{6+}$ ions changes. It was determined that a higher removal percentage of uranium is obtained at the beginning of the adsorption. $K_d$ and percentage sorption of uranium at the optimum adsorption conditions were found as $196 \pm 6$ m $Lg^{-1}$ and $66.2 \pm 0.7\%$, respectively. The results show that TAB samples can be an alternative low-cost adsorbent for removing $U^{6+}$ ions from aqueous solutions.

Mishra and Patel [42] studied the use of activated carbon, kaolin, bentonite, blast furnace slag, and fly ash as adsorbent with a particle size between 100 mesh and 200 mesh to remove the lead and zinc ions from water. The concentration of the solutions prepared was in the range of 50–100 mg $L^{-1}$ for lead and zinc for single and binary systems which were diluted as required for batch experiments. The effect of contact time, pH, and adsorbent dosage on the removal of lead and zinc by adsorption was investigated. The equilibrium time was found to be 30 minutes for activated carbon and 3 hours for kaolin, bentonite, blast furnace slag, and fly ash. The most effective pH value for lead and zinc removal was 6 for activated carbon. Variation in pH value did not affect lead and zinc removal significantly for other adsorbents. Adsorbent doses were varied from 5 gL$^{-1}$ to 20 gL$^{-1}$ for both lead and zinc solutions. An increase in adsorbent doses increased the percent removal of lead and zinc. A series of isotherm studies were undertaken and the data evaluated for compliance was found to match with the Langmuir and Freundlich isotherm models. To investigate the adsorption mechanism, the kinetic models were tested, and it follows second-order kinetics. Kinetic studies revealed that blast furnace slag was not effective for lead and zinc removal. From the results, it was found that bentonite and fly ash were effective for lead and zinc removal. Batch tests were carried out by Yuan and his group [43] to investigate the removal mechanism of hexavalent chromium [$Cr^{6+}$] by montmorillonite-supported magnetite nanoparticles. Montmorillonite-supported magnetite nanoparticles were prepared by coprecipitation and hydrosol method [44]. The obtained materials were characterized by X-ray diffraction, nitrogen adsorption, elemental analysis, differential scanning calorimetry, transmission electron microscopy, and X-ray photoelectron spectroscopy. The average sizes of the magnetite nanoparticles without and with montmorillonite support were found to be around 25 and 15 nm, respectively. The montmorillonite-supported magnetite nanoparticles were present on the surface or inside the interparticle pores of clays, with better dispersing and less coaggregation than the ones without montmorillonite support. The $Cr^{6+}$ uptake were mainly governed by a physicochemical process, which included an electrostatic attraction followed by a redox process in which $Cr^{6+}$ was reduced into trivalent chromium. The adsorption of $Cr^{6+}$ was highly pH dependent, and the kinetics of the adsorption followed the pseudosecond-order model. The adsorption data of unsupported and clay-supported magnetite nanoparticles followed Langmuir and Freundlich isotherm models. The montmorillonite-supported magnetite nanoparticles showed a much better adsorption capacity per unit mass of magnetite (15.3 mg g$^{-1}$) than unsupported magnetite (10.6 mg g$^{-1}$) and were more thermally stable than their unsupported counterparts.

Cobalt adsorption efficiencies of kaolinite and montmorillonite was studied by Angove et al. [45] and Bhattacharyya and Sen Gupta [11]. Angove et al. [45] found the langmuir capacities as 1.5 mg g$^{-1}$ at 313 k and Bhattacharyya and Sen Gupta found it to 11.2 mg g$^{-1}$ for kaolinite. For montmorillonite, Bhattacharyya and Sen Gupta found the Langmuir and freundilich values to be 28.6 and 4.6, respectively. Yavuz et al. [29] showed that adsorption of $Cu^{2+}$ on kaolinite followed langmuir isotherm with adsorption capacity equal to 11.0 mg g$^{-1}$. Pandey and Mishra [16] prepared organic-inorganic composite of chitosan and nanoclay (Cloisite 10A) with combined properties of hydrophilicity of an organic polycation and adsorption capacity of inorganic polyanion. The chitosan/clay nanocomposite (CCN) was prepared by solvent casting method. The chemical, structural, and textural characteristics of the material were determined by FTIR, XRD, TEM, SEM, and EDAX analysis. XRD and TEM results indicated that an exfoliated structure was formed with addition of small amounts of MMT-Na$^+$ (montmorillonite-Na$^+$) to the chitosan matrix. These composite material were used for the removal of $Cr^{6+}$ from aqueous solution. They showed that pH$_3$ was found most suitable and adsorption data fits the Langmuir and Freundlich isotherms. The adsorption showed pseudosecond order kinetics with a rate constant of 8.0763 $*10^{-4}$ g mg$^{-1}$ min$^{-1}$ at 100 ppm $Cr^{6+}$ concentration. The natural kaolin calcined at 550°C (mostly meta kaolin) and raw bentonite (mostly montmorillonite) pretreated with $Fe^{2+}$, $Fe^{3+}$, $Al^{3+}$, and $Mn^{2+}$ salts were used to remove As from the model anoxic groundwater with As$^{3+}$concentration about 0.5 and 10 mg L$^{-1}$ by Doušová et al. [46]. The efficiency of As$^{3+}$ sorption varied from 92% to >99% by the sorption capacity higher than 4.5 mg g$^{-1}$. In the case of metakaolin, $Fe^{2+}$ and $Mn^{2+}$ treatments proved the high sorption efficiency > 97%), while only <50% of As was removed after $Fe^{2+}$and $Al^{3+}$ pretreatment. The sorption capacities of treated metakaolin ranged from 0.1 to 2.0 mg g$^{-1}$. The utilization of low-grade clay materials as selective sorbents is one of the most effective possibilities of As removal from contaminated water reservoirs. It was shown that simple pretreatment of these materials with Fe (Al and Mn) salts significantly improved their sorption affinity to As oxyanions. While many treatment technologies are available for arsenic removal from drinking water including coagulation/filtration, lime softening, activated alumina adsorption, ion exchange, and membrane processes, most of these approaches are expensive and more suitable for large water systems. In this study, membranes made of low-cost clay minerals were explored by Fang et al. [47] for arsenate removal. Montmorillonite, kaolinite, and illite were selected for membrane preparation. Feed water spiked with arsenate was pumped through the compacted clay membranes and the effluent was collected at the lower pressure side for arsenic analysis. The ability of clay membranes to retain arsenic was investigated at different initial arsenic concentrations and ionic strengths controlled by sodium chloride. The influence of applied pressure and the permeate flux on arsenic removal efficiency was also examined. The results indicated that a greater than 90% of arsenic rejection could be achieved for water with 50–100 $\mu g\,L^{-1}$ of arsenate using the clay membranes. It was observed that the required pressure for clay membrane filtration was significantly higher than that of synthetic organic membranes.

Water and wastewater studies in Malawi, South Africa, revealed very high levels of heavy metals in most streams and other water bodies particularly within their urban areas. The metals are produced and released during industrial and agricultural activities and also in vehicular emissions. The study conducted by Sajidu et al. [48] investigated the potential of mixed clay, obtained from the Tundulu area in removing $Cd^{2+}$, $Cr^{3+}$, $Cu^{2+}$, $Hg^{2+}$, $Pb^{2+}$, and $Zn^{2+}$ cations and anions from aqueous solutions using batch equilibrium technique. Qualitative mineralogical characterization of the clay revealed that the clay contains illite, distorted kaolinite, mixed layer minerals, and nonclay mineral carbonate fluoroapatite. pH pzc for the raw clay, as determined by potentiometric titrations, was 9.66, while pH pzc of pretreated clay was 9.63. Pretreatment of the clay involved removal of carbonates, iron oxides, and organic matter. Initial total metal concentrations ranged from 3 to 5 mg $L^{-1}$. pH metal sorption dependence of the clay revealed $Cr^{3+}$ removal from pH of 3 to complete removal at pH 5 with over 90% of the removal attributable to adsorption on the clay while the remaining 10% attributable to both adsorption, and $Cr(OH)_3$ precipitation. $Zn^{2+}$ complete removal occurred at pH above 7 with 92% attributable to adsorption while the rest could be from both adsorption and hydroxide precipitation. $Cu^{2+}$ was removed from pH 4 and completely above pH 6.8 with 50% due to adsorption. $Cd^{2+}$ removal was between pH of 6 and 9 with 85% due to adsorption to the clay. Lead was completely removed at pH greater than 7.67. Removal of $Hg^{2+}$ at total $Hg^{2+}$ concentration of 0.023 mM was pH independent fluctuating between 30% and 60%. No effective removal of $AsO_4^{-3}$ anion was observed.

Selenium is a natural trace element found in bedrock, but it is also introduced into the environment by anthropogenic activities, such as mining and combustion of fossil fuels [49, 50]. At low concentrations, selenium is an essential micronutrient for mammals, but consumption of quantities exceeding daily recommendations can cause health problems. Its toxicity [51] led the World Health Organization (WHO) and the European Union (EU) to recommend a maximum selenium concentration in drinking water of 10 ppb, while the Environment Protection Agency (EPA) sets a limit of 50 ppb. Chitosan-montmorillonite composites were designed by Bleiman and Mishael [52] to adsorb selenium from water and its efficiency was compared with widely used commercial adsorbent aluminum oxide (Al-oxide) and feric oxide (Fe-oxide). The highest adsorption efficiency was obtained for chitosan-montmorillonite composites. These composites were characterized by XRD, zeta potential, and FTIR measurements. Adsorption isotherms of selenate on the composite, on Al-oxide, and on Fe-oxide were in good agreement with the Langmuir model, yielding a somewhat higher capacity for the composite, 18.4, 17.2, and 8.2 mg g$^{-1}$, respectively. In addition, adsorption by the composite was not pH dependent, while its adsorption by the oxides decreased at higher pH. Selenium removal from well water (closed due to high selenium concentrations, 0.1 mg L$^{-1}$) by the composite, brought levels to below the WHO limit (0.01 mgL$^{-1}$) and was selective for selenium even in the presence of sulfur (13 mg L$^{-1}$). Selenium adsorption by the composite was higher than by the Al-oxide due to high adsorption of sulfur by the latter. It was found that the polymer clay composite is an innovative sorbent that in suspension efficiently removed selenium from well water with high selenium concentrations. On the other hand, the Al-oxide efficiently removed selenium from the well water when employed in filtration columns. Na et al. [53] investigated the adsorption of arsenate and arsenite from aqueous solutions using Ti-pillared montmorillonite (Ti-MMT) as a function of contact time, pH, temperature, coexisting ions, and ionic strength. The adsorption of both arsenate and arsenite were temperature and pH dependent, indicating different adsorption mechanisms. The effect of coexisting ions on the adsorption was also studied, and, among the ions investigated, only phosphate had a noticeable influence on the adsorption of arsenate, while the effect of other ions was negligible. A pseudosecond-order chemical reaction model was obtained for both arsenate and arsenite;

adsorption isotherms of arsenate and arsenite fitted the Langmuir and Freundlich isotherm models well. X-ray diffraction (XRD) and X-ray photoelectron spectroscopy (XPS) were used to study the nature of surface elements before and after adsorption. This work demonstrates that Ti-pillared montmorillonite is an efficient material for the removal of arsenate and arsenite from aqueous solutions. It was found that the removal capacity of arsenite decreased with increase in temperature, whereas the adsorption of arsenate shows the same trend in the temperature range 25°C –35°C but an inverse trend between 45°C and 65°C. To test the effect of coexisting anions (phosphate, nitrate, and sulfate) present in natural water on arsenite and arsenate removal, simulated water experiments were performed. Experimental results showed that none of these three anions had any noticeable influence on the adsorption of arsenite, while the presence of phosphate significantly decreased the adsorption capacity of arsenate. The reason for decrease in absorption efficiency of $As^{5+}$ in the presence of phosphate was explained as competition between phosphate and arsenate ions on limited availability of adsorption sites. Ake et al. [54] prepared a binary clay composite for adsorption of lead from the contaminated water. They prepared clay-carboxy methyl cellulose composite. The binary composite had higher efficiency for adsorption of lead from solution, while providing void volume, increased surface area, and considerably enhanced hydraulic conductivity. The results suggested that a combination of sodium montmorillonite clay and carbon exhibited enhanced sorption of lead compared to carbon alone and also supported the potential application of various combinations of sorbent materials. Zhao et al. [55] prepared clay composite by embedding bentonite in polyacrylamide (PAAm) gels for adsorption of $Cu^{2+}$ ions from the ground and surface waters. These composite materials combine the elasticity and permeability of gels with the high ability of clays to adsorb heavy metal ions. The sorption and desorption of $Cu^{2+}$ on bentonite-polyacrylamide (BENT-PAAm) was investigated by Zhao group as the function of pH, ionic strength, adsorbent content, Cu(II) concentrations, and temperature. The results indicated that the sorption of $Cu^{2+}$ on BENT-PAAm was strongly dependent on pH, ionic strength, and temperature. The sorption increased from about 9% to 97% at pH ranging from 2.4 to 7. The sorption of $Cu^{2+}$ on BENT-PAAm increased with increasing temperature and decreasing ionic strength. The sorption of $Cu^{2+}$ on BENT and on

BENT-PAAm was reported as endothermic and irreversible process. A superabsorbent composite (SAC) was synthesized by copolymerization reaction of partially neutralized acrylic acid (AA) on bentonite micropowder using N, N_-methylenebisacrylamide as a crosslinker and ammonium persulfate as an initiator in aqueous solution by Bulut et al. [56] using solution polymerization technique. The water absorbency of synthesized SAC was reported as 352 and 110 g $H_2Og^{-1}$ in distilled water and 0.2% NaCl, respectively. Sorption capacity of SAC was investigated for heavy metal ions (HMI) using Langmuir and Freundlich model of adsorption. The maximum adsorption capacity of HMI onto the bentonite-based SAC from their solution was found to be 1666.67, 270.27, 416.67, and 222.22 mg g$^{-1}$ for $Pb^{2+}$, $Ni^{2+}$, $Cd^{2+}$, and $Cu^{2+}$, respectively.

# Inorganic Contaminant

Inorganic contamination of drinking water and its sources is caused by natural and anthropogenic factors. Fluorosis is endemic in at least 25 countries across the globe and has affected millions of people [75–77]. It is caused by high concentration of fluoride above 1.5 mg L$^{-1}$ in drinking water [78]. Fluoride is beneficial when present within the permissible limit of 1.0–1.5 mg L$^{-1}$ for calcification of dental enamels [79]. Similarly, excess of nitrates in drinking water causes methemoglobinemia or blue baby disease. Thus, management of these inorganic contaminants in water is of prime importance. Low-cost bentonite clay was chemically modified using magnesium chloride in order to enhance its fluoride removal capacity by Thakre et al. [57]. The magnesium incorporated bentonite (MB) was characterized by using XRD and SEM techniques. Batch adsorption experiments were conducted to study and optimize various operational parameters such as adsorbent dose, contact time, pH, effect of coions, and initial fluoride concentration. It was observed that the MB works effectively over wide range of pH and showed a maximum fluoride removal capacity of 2.26 mg g$^{-1}$ at an initial fluoride concentration of 5 mg L$^{-1}$, which is much better than the unmodified bentonite. The experimental data fit well into Langmuir adsorption isotherm and followed pseudofirst-order kinetics. Thermodynamic study suggested that fluoride adsorption on MB is reasonably spontaneous and an endothermic process. MB showed significantly high fluoride removal in synthetic water as compared to field water. Desorption study of MB suggested that almost all the loaded fluoride was desorbed (~97%) using 1 M NaOH solution; however, maximum fluoride removal decreased from 95.47 to 73 (%) after regeneration. From the experimental results, it was inferred that chemical modification enhances the fluoride removal efficiency of bentonite and it works as an effective adsorbent for defluoridation of water.Adsorption potential of metal oxide (lanthanum (La), magnesium (Mg), and manganese (Mn)) incorporated bentonite clay was investigated by Kamble et al. [58] for defluoridation of drinking water using batch equilibrium experiments to gain insight of adsorption behavior, kinetics, and mechanisms of adsorption of fluoride ion. The effect of various physicochemical parameters such as pH, adsorbent dose, initial fluoride concentration, and the presence of interfering coions on adsorption of fluoride has been investigated. The 10% La-bentonite showed higher fluoride uptake capacity for defluoridation of drinking water as compared to Mg-bentonite, Mn-bentonite, and bare bentonite clay. The uptake of fluoride in acidic pH was higher as compared to alkaline pH. The equilibrium adsorption data fitted reasonably well in both Langmuir and Freundlich isotherm models. It was also observed that the presence of certain coexisting ions can have positive effect on removal of fluoride, while carbonate and bicarbonate anions showed deleterious effect. The rate of adsorption was reasonably rapid, and maximum fluoride uptake was attained within 30 minutes. The modified adsorbent material showed better fluoride removal properties for actual field water, which could be due to the positive effect of other coions present in the field water.Defluoridation from aqueous solutions by zirconium- (Zr) loaded bentonite (ZLB) was studied by Ma et al. [59]. It has been found that maximum adsorption of fluoride from aqueous solutions takes place below pH 6. The fluoride adsorption followed the mechanism of ion exchange.Groundwater pollution by nitrates is a widespread problem in many locations in the world. The underground aquatic mantle of the Peninsula of Yucatan, Mexico, is highly vulnerable due to its karstic nature. Adsorption methods are a good choice for nitrate elimination. In this work, Mena-Duran et al. [60] modified natural calcium bentonite by acid thermoactivation using HCl and $H_2SO_4$, and tested it as a media for nitrate removal in an aqueous solution. Acid thermoactivation is a process that allows the controlled extraction of aluminum ions from the crystalline structure of the clay, also introducing an acceptable acidity level. This process modifies the textural properties of clays favoring a greater porosity and surface area. The nitrate concentration in the solution was measured by FT-IR, using the Lambert-Beer law. Clay characterization was carried out by X-ray diffraction and FT-IR spectroscopy; surface area was measured by the BET method. It was shown by Mena-Duran and group that natural clays modified by acid thermoactivation are capable of nitrate removal in aqueous solutions. Nitrate adsorption was found to be proportional to the stirring time. Calcium montmorillonite activated by hydrochloric acid showed a better nitrate removal capacity, up to 22.28% when the stirring time was increased from 0.5 hours to 68 hours. The ionic exchange was confirmed by the presence of KCl in the clay residue. The BET area measurements showed no direct relation between the surface area and the nitrate removal capacity. Infrared spectroscopy was used to measure the nitrates adsorbed by the natural clay materials.

# Organic Contaminants

Clays offer an attractive and inexpensive option for the removal of organic and inorganic contaminants [80]. The adsorption of several organic contaminants in water such as pesticides, phenols, and chlorophenols has been reported recently in the literature [81–87].Chloroacetic acids, such as trichloroacetic acid (TCAA), dichloroacetic acid (DCAA), and monochloroacetic acid (MCAA) are receiving increasing attention in the literature [88]. These chloroacetic acids, commonly formed during the reactions between chlorine and natural organic matter (NOM) during prechlorination or disinfection in potable water production, have been shown to be carcinogenic and may potentially pose a risk to human health [89].Chloroacetic acids, formed during the disinfection process in potable water production, are considered to pose a potential risk to human health. In this work, Gu et al. [61] investigated dichloroacetic acid (DCAA) removal from drinking water by using a process of bentonite-based adsorptive ozonation. This process is formed by combined addition of ozone, bentonite, and $Fe^{3+}$. During the reaction, DCAA is removed by the joint effect of adsorption, ozonation, and catalytic oxidation. In addition, under the effect of the adsorption, natural organic matters (NOM) can be adsorbed onto the bentonite surface, resulting in a reduced scavenging effect toward HO·radicals, and hence eliminate the negative effect of NOM on DCAA removal.

It was explained that at the initial stage of the reaction, $Fe^{3+}$ is rapidly hydrolyzed to polycations and adsorbed onto the bentonite surface or into its structural layers. This charges the surface of the bentonite with positive ions, which, in turn, increases its surface area, resulting in a strong adsorption of HA or DCAA. Furthermore, $Fe^{3+}$ catalyzes ozone decomposition to form HO·, thus further improving the efficiency. Experimental results show that ozonation alone removes 40%, bentonite (dosage 100 mg $L^{-1}$) enhanced ozonation removes 51%, and addition of $Fe^{3+}$ increases the DCAA removal by 68%. It was observed that increasing the dose of $Fe^{3+}$ from 0.5–5 mg $L^{-1}$, DCAA removal increases significantly to 92%. The optimum time was found to be 40 minutes. The adsorptive ozonation has been shown to be potentially advantageous in destruction of toxic, dissolved pollutants in drinking water and appears to have great potential for a wide range of treatment applications.In another study by Lu and Pan [62], the adsorption method using quaternary ammonium salt modified bentonite as the adsorbent was used for the removal of carbon tetrachloride (CT) in simulated groundwater sample. The morphology of bentonite before and after modification was observed through scanning electron microscope (SEM). The effects of bentonite dosage, adsorption time, and temperature on removal rate of CT were investigated. The SEM results showed that after natural calcium-bentonite is modified by quaternary ammonium salt, the particle clearance of the modified bentonite obviously increased, and the surface property changed from originally hydrophilic to hydrophobic, enabling it to effectively adsorb and remove CT from water. The optimal adsorption conditions obtained were bentonite dosage 5.0 g$L^{-1}$, adsorption time 2 hours, and temperature 30°C. Under these optimized conditions, the removal efficiency of CT reached above 70% when modified bentonite was used in treating actual groundwater polluted by CT.Rivera-Jimanez [63] modified inorganic-organic-intercalated (IO) bentonites with $Co^{2+}$, $Ni^{2+}$, or $Cu^{2+}$ and created adsorbents for the removal of relevant emerging contaminants (naproxen, salicylic acid, clofibric acid, and carbamazepine) from water to overcome challenges associated with low concentration and polar nature of these contaminants. Characterization of the materials using X-ray diffraction, porosimetry, scanning electron microscopy, thermal gravimetric analysis, and Fourier transform infrared spectroscopy indicated general structural integrity. It was found that the metal loading increased in the following order:

Ni < Cu < Co. Single-point adsorption experiments were done at room temperature with different pH conditions and using initial adsorbate concentration of 14 ppm. In general, the transition metal-modified IO bentonites displayed adsorption capacities that varied depending on the type of metal, pH, and nature of the adsorbate. The largest adsorption capacity was observed for salicylic acid, with removal of $5.5\,\mu molg^{-1}$ using $Cu^{2+}$ modified hexadecyltrimethylammonium bromide (HDTMAB, ≥99.0% purity) natural bentonites. According to Rivera-Jimenez, the behavior could have been because of its smaller footprint. The higher pH was found to be the optimum for maximum removal efficiency. From the results, it was concluded that the presence of some functional groups plays an important role during the adsorption of a particular adsorbate, possibly indicating complexation with the transition metal. For carbamazepine, although the observed adsorption loadings were comparable to those of other adsorbents, the modification of the IO bentonites does not appear to enhance the unmodified material capacity. This could have been because of the absence of key functional groups in this particular adsorbate.

A natural bentonite modified with a cationic surfactant, acetyl trimethylammonium bromide (CTAB), was used as an adsorbent for removal of phenol from aqueous solutions by Senturk et al. [64]. The natural and modified bentonites (organobentonite) were characterized by FTIR, XRD, and SEM. Batch adsorption experiments were performed to study the effects of various parameters such as solution pH, contact time, initial phenol concentration, organobentonite concentration, and temperature on phenol adsorption onto organobentonite. Maximum phenol removal was observed at pH 9.0. Equilibrium was attained after contact of one hour only. At equilibrium, the organobentonite concentration was $10\,gL^{-1}$ with initial phenol concentration of $100\,mg\,L^{-1}$. The adsorption isotherms were described by Langmuir and Freundlich isotherm models, and both models fit well. The monolayer adsorption capacity of organobentonite was found to be $333\,mg\,g^{-1}$. Combined ozonation and bentonite coagulation process (COBC) was investigated by Gu et al. [65] as a method of concurrently removing humic acid (HA) and o-dichlorobenzene (DCB) from drinking water. When compared with only ozonation and coagulation, COBC was highly efficient in removing the HA and DCB concurrently. In this study, it was shown that in this process, HA and DCB were removed by joint effect of catalytic ozonation and bentonite coagulation. HA removal was highly dependent on the coagulation process, while DCB removal was dependent on the oxidation process in COBC. Iron in solution not only acted as a coagulant, but also promoted the formation of HO·, which is effective in destroying aromatic chemicals. Bentonite in COBC improved the coagulation process, resulting in enhancement in the treating efficiency. COBC has proved to be potentially advantageous on dissolved pollutants in drinking water and appears to have great potential for a wide range of practical applications. Algae are most common aquatic species which grow and inhabit sea, lakes, and rivers. Very rapid growth of algae can, therefore, cause problems in the water supply industry, such as water discoloration, taste, odor, and blockage of filters. In particular, some types of algae (e.g., blue-green algae) can be toxic to humans and other organisms. If an excessive growth of algae occurs, endogeneous toxins (e.g., microcystin and nodularin) emitted act as hepatoxins, while anatoxin and saxitoxin act as neurotoxins; these may cause serious damage to both humans and animals alike [90, 91]. Therefore, the blooming of algae not only worsens water treatment performance and then deteriorates water quality, but also results in the toxic effect on human beings and animals.In this study, Jiang and Kim [66] compared the algae removal efficiency of clay composites with widely used commercial metal coagulants in simulated water. The coagulants used in this study included aluminium sulphate (AS) and polyaluminium chloride (PACl). AS had the Al content of 8% w/w as $Al_2O_3$ and PACl (PAX-XL9) had the Al content of 8.5% w/w as $Al_2O_3$, with the basicity of 53.8, and four clays, bentonite (Bent), sodium modified bentonite (Na-Bent), montmorillonite KSF (Mont-KSF), and montmorillonite K10 (Mont-K10). A standard jar test procedure [92] was applied to assess the coagulation performance. The shape of algae-clay flocs was examined by a scanning electron microscope (SEM). Experimental results show that coagulation with both metal coagulants and clays effectively removed algal cells from water. Prepolymerized inorganic coagulants (e.g., polyaluminium chloride, PACl) normally contain high positive charge so algal removal occurs by charge neutralization mechanism. This results in better performance of PACl in the removal of algae. In comparison with aluminium sulphate (AS), polyaluminium chloride (PACl) achieved a better performance with respect to algal removal. Excellent algal removal efficiency of the clays Bent and Mont-KSF was observed. For a dose of 200 mg/L, the percentage removal ofchlorophyll-a achieved by both clays was 100%, which is greater than the efficiency of PACl, for all the dose ranges investigated. In turbidity reduction experiments, Bent did not perform as well as Mont-KSF. The doses clay and PACl for removal of algae was found to be $200 \, mg \, L^{-1}$ and $1.5 \, mg \, L^{-1}$, respectively. Abundant availability, low cost, together with nonharmful chemical residuals in the treated water by clays make it a very efficient alternative treatment reagent in coping with the problems of algae. In another study by Gao et al. [67], montmorillonite-Cu (II)/Fe(III) oxides magnetic material was prepared and used for removal of harmful algae from water. The material was prepared by the support of $Cu^{2+}/Fe^{3+}$ oxides on pillared montmorillonite and was characterized by XRD, zeta potential measurements. The prepared magnetic material was effective for the removal of cyanobacterial Microcystis aeruginosa, and the loaded particles were chemically regenerated using acetone solution. The removal increased with the decrease in pH and the increase of ionic strength and $Ca^{2+}$ concentration.

Atrazine (2-chloro-4-ethylamino-6-isopropylamino-s-triazine) is currently one of the most widely applied herbicides in the US [93]. It has been detected at high concentrations in ground and surface waters all over Europe and North America [94–97] due to its extensive use, ability to persist in soils, low sediment partitioning, slow rate of degradation, and its tendency to travel with water. In the US, the upper limit for atrazine in drinking water is 3 ppb, whereas the European Union legislation banned its use (since 2003) and fixed a limit of 0.1 ppb [98]. Several methods and technology had been used to remove atrazine from contaminated water. But all of them proved to very expensive. A number of studies explored modifying silicate minerals (clays and zeolites) as adsorbents for atrazine. Lemi et al. [99] examined removal of atrazine, lindane, and diazinon from water by organozeolites. However, the adsorption capacity for atrazine was the lowest ($2.0 \, mmolg^{-1}$). Sanchez- Martin et al. [100] employed clay minerals modified by a cationic surfactant in batch and column experiments for the adsorption of pesticides. The results showed an increase in the adsorption ability compared to untreated clays, but the adsorption coefficient of atrazine was very low. Borisover et al. [101] reached high sorption values of atrazine (98% after 18 h) on dye-clay complexes; however, the adsorbent concentration was very high ($50 \, gL^{-1}$). Modification of the clay minerals vermiculite and montmorillonite by intercalating $F_e(III)$ polymers for the removal of atrazine and its metabolites was examined [102]. Enhancement in the adsorption capacity was observed for both intercalated clays in comparison with the potassium clays although the adsorption time was relatively high (24 h). The studies described above did not include the effect of dissolved organic matter (DOM) or a comparison to granular activated carbon (GAC). Streat and Sweetland [103] compared the adsorption of atrazine by hypersol-macronet polymer phases and by GAC F-400 and found that the GAC was more effective for the removal of atrazine than the polymer phases. Only a few studies focused on applying polymer-clay composites as sorbents for organic pollutants [104, 105]. The study of Churchman [104] on the formation of polycation-clay composites and their use as sorbents for nonionic and anionic pollutants. Breen [105] examined the use of polycation-exchanged clays as sorbents for organic pollutants and studied the influence of layer charge on pollutant sorption capacity. Radian and Mishael [106] studied the binding of an herbicide to polycation-clay composites for the design of controlled release herbicide formulations. Zadaka et al. [68] studied the effect of DOM on atrazine removal by the polymer-clay composite and by GAC. The removal of atrazine (2-chloro-4-ethylamino-6-isopropylamino-s-triazine), a widely used herbicide, removal from water by two polycations preadsorbed on montmorillonite was studied. Batch experiments demonstrated that the most suitable composite poly (4-vinylpyridine-co-styrene)-montmorillonite (PVP-co-S-90%-mont.) removed 90–99% of atrazine (0.5–28 ppm) within 20–40 min at 0.367% w/w. Calculations employing Langmuir's equation could simulate and predict the kinetics and final extents of atrazine adsorption. Columns filter experiments (columns 20 × 1.6 cm) which included 2 g of the PVP-co-S-90%-mont. composite mixed with excess sand removed 93%–96% of atrazine (800 ppb) for the first 800 pore volumes, whereas the same amount of granular activated carbon (GAC) removed 83%–75%. In the presence of dissolved organic matter (DOM; 3.7 ppm), the efficiency of the GAC filter to remove atrazine decreased significantly (68%–52% removal), whereas the corresponding efficiency of the PVP-co-S-90%-mont. filter was only slightly influenced by DOM. At lower atrazine concentration (7 ppb), the PVP-co-S-90%-mont. filter reduced even after 3000 pore volumes the emerging atrazine concentration below 3 ppb (USEPA standard). In the case of the GAC filter, the emerging atrazine concentration was between 2.4 and 5.3 $\mu g L^{-1}$ even for the first 100 pore volumes. From the above experimental results, the PVP-co-S-90%-mont. composite can be used as an efficient material for the removal of atrazine from water. This study by Undabeytia et al. [69] presents the vesicle-clay complex as a powerful sorbing material for water purification of organic contaminants by both filtration and sedimentation. A main advantage of the vesicle-clay system stems from providing a relatively large number of highly hydrophobic sites, which yields efficient and large capacity of adsorption for neutral and anionic pollutants. Vesicle-clay complexes in which positively charged vesicles composed of didodecyldimethylammonium bromide (DDAB) were adsorbed on montmorillonite removed efficiently anionic (sulfentrazone and imazaquin) and neutral (alachlor and atrazine) pollutants from water. These complexes (0.5% w:w) removed 92%–100% of sulfentrazone, imazaquin, and alachlor and 60% of atrazine from a solution containing $10\,mgL^{-1}$ of it. A synergistic effect on the adsorption of atrazine was observed when all pollutants were present simultaneously ($30\,mg\,L^{-1}$ each), its percentage of removal being 85.5 percent. Column filters (18 cm) filled with a mixture of quartz sand and vesicle-clay (100 : 1, w : w) were tested. For the passage of one liter (25 pore volumes) of a solution including all the pollutants at $10\,mg\,L^{-1}$ each, removal was complete for sulfentrazone and imazaquin, 94% for alachlor, and 53.1% for atrazine, whereas removal was significantly less efficient when using activated carbon. The research by Rytwo et al. [70] show the efficient sorption of naphthalene and several phenolic derivatives to organoclays prepared by adsorption of crystal violet or tetraphenylphosphonium ions on montmorillonite until a charged-neutralized surface is obtained. The amounts of pollutant adsorbed are at least of the same order of magnitude as those measured for high-quality activated carbon, but the adsorption proceeds almost immediately, whereas for activated carbon it takes longer. The proposed organoclays were mixed with sand and tested in column filters, showing complete removal of high concentration of pollutant at several pore volumes. The adsorbents can also efficiently be applied in sequential batch reactors due to the fast adsorption kinetic, followed by flocculation that allows easy separation of the purified effluent. A volume of 150 mL of a 1000 µM TCP solution was completely purified to levels below 3 µM, by means of 0.25 g organoclay. The work by Bonina et al. [71] described the interactions between two commercial clays, bentonite and kaolin, and an iron-salicylate complex on the removal efficiency of salicylic acid. Adsorption experiments were accomplished using a water solution containing $Fe^{3+}$ 0.0176 M and salicylic acid 0.0253 M. Natural and treated clay samples were characterized by chemical analyses, powder X-ray diffraction and thermal analyses. It was shown that time dependence of salicylic acid adsorption by bentonite followed first-order kinetics, with respect to the percentage of salicylic acid adsorbed, in the first twelve hours; afterwards, the reaction slowed down. The reaction is completely exhausted after 2 days, and during the next 4 days, the concentration of salicylic acid in bentonite does not change from its asymptotic value of 8.0%. The adsorption kinetics of salicylic acid by kaolin showed a slow adsorption beginning after the fourth day of treatment and finished after 19 days. The amount of salicylic acid adsorbed was 5.5% of the final complex. The release of salicylic acid adsorbed by bentonite and kaolin was tested in 0.2 N solutions of $Na^+$, $K^+$, $Mg^{2+}$, and $Ca^{2+}$. Salicylic acid release rates from $F_e(III)$-salicylate-containing bentonite were also measured through cellulose acetate membranes by means of Franz-type diffusion cells: an initial slow release of salicylic acid was followed by a fast release phase; after 23 hours, the concentration of salicylic acid released can be considered constant and the drug desorbed was 1.4% of the amount adsorbed by the bentonite. Even if the desorbed amount of salicylic acid is not very high, the bentonite-salicylate complex could be suitable for an application by gradual release. Wang et al. [107] prepared a series of cetyltrimethyl ammonium bromide bentonite (C-Bt), cationic polyacrylamide bentonite (P-Bt), and composite organobentonite (C/P-Bt) by modifying bentonite with cetyltrimethyl ammonium bromide (CTMAB) and/or cationic polyacrylamide (CPAM). They measured the basal spacings of the synthesized organobentonites using XRD. The sorption capacities of phenol and nitrobenzene to these organobentonites from water were examined. The results showed that the basal spacing values of C/P-Bt were larger than those of C-Bt and P-Bt, which indicated a simultaneous intercalation of bentonite interlayers by CTMAB and CPAM. The sorption capacity of C/P-Bt was better than that of C-Bt. Under the same equilibrium concentration (7045 mg $L^{-1}$ for phenol and 409 mg $L^{-1}$ for nitrobenzene), the sorbed amounts of phenol and nitrobenzene on 60C/4%P-Bt were 150 and 69 mg $g^{-1}$, which enhanced 26% and 28%, respectively, comparing with those on 60C-Bt. There was an improved adsorption efficiency because of the arrangement model of CTMAB within the C/P-Bt interlayers was affected by CPAM, which led to the formation of organic phase with better affinity to the organic compounds.

Carbamazepine is a prescription anticonvulsant and mood stabilizing pharmaceutical administered to humans. Carbamazepine is persistent in the environment and frequently detected in water systems. Sorption and desorption of carbamazepine from water was measured for smectite clays with the surface negative charges compensated with $K^+$, $Ca^{2+}$, $NH_4^+$, tetramethylammonium (TMA), trimethylphenylammonium (TMPA), and hexadecyltrimethylammonium (HDTMA) cations by Zhang et al. [73]. Sorption of carbamazepine by TMPA and HDTMA smectites from water was found to be 10–150 times greater than sorption by K-, Ca-, and TMA-smectites within the measured aqueous carbamazepine concentrations (0.1–1.0 mg $L^{-1}$). As per the experimental results, the magnitude of sorption followed the order: TMPA-smectite ≥ HDTMA-smectite > NH4-smectite > K-smectite > Ca-smectite ≥ TMA-smectite. The greatest sorption of carbamazepine by TMPA-smectite is attributed to the interaction of conjugate aromatic moiety in carbamazepine with the phenyl ring in TMPA through ϖ-ϖinteraction. Partitioning process is the primary mechanism for carbamazepine uptake by HDTMA smectite. For $NH_4$-smectite, the urea moiety in carbamazepine interacts with exchanged cation by H-bonding, hence demonstrating relatively higher adsorption. Sorption by K-, Ca-, and TMA-smectites from water occurs on aluminosilicate mineral surfaces [73].

# Pathogens

The microcystins are potent mammalian liver toxins [107], known to be potent and specific in vitro inhibitors of the catalytic subunits of protein phosphatases-1 and 2A [108, 109] and are extremely potent tumor promoters [110, 111]. Since it is widely suspected that many conventional water treatment methods are ineffective at reducing human exposure to microcystins, investigations into economic and practical methods of remedial water treatment are important. Although removal of cyanobacterial cells and toxins from drinking water using domestic water filters resulted in marginal success [112], it is following the termination of cyanobacterial growth that the majority of microcystins are considered to enter into the surrounding water after lysis and cell death. To date, perhaps photoirradiation is the most promising new method for detoxifying microcystins in raw water [113–115]. In a recent report Harada and Tsuji [116] looked at the persistence and decomposition of these hepatotoxins in the natural environment. Five pathways were considered as contributing to natural routes of detoxification. Of relevance to the work presented here is that it was ascertained that microcystins are absorbed strongly on sediment and that they are difficult to recover. The results of microcystin-LR scavenged by naturally occurring clay minerals are reported by Harada and Tsuji. The microcystin cyanobacterial hepatotoxins represent an increasingly severe global health hazard. Since microcystins are found worldwide in drinking water reservoirs concern about the impact on human health has prompted investigations into remedial water treatment methods. The preliminary study by Morris et al. [74] investigated the scavenging from water of microcystin-LR by fine-grained particles known to have a high concentration of the clay minerals kaolinite and montmorillonite. The results show that more than 81% of microcystin-LR can be removed from water by clay material. Thus, microcystin-LR is indeed scavenged from water bodies by fine-grained particles and that this property may offer an effective method of stripping these toxins from drinking water supplies.

# CONCLUSIONS

Table 1 summarizes the variety of pollutants treated with different types of clays, their efficiency, and the effect of different variables on their adsorption capacity. From the table, it is clear that natural clay and its composites are capable of removing contaminants ranging from metals to priority pollutants from contaminated drinking water and its sources. Results from the recent advances in using natural clay and its modified composites show the flexible nature of the clay and its ecofriendly nature. They are capable of removing organic and inorganic contaminants from drinking water with very high removal ratios of toxic trace metals, nutrients, and organic matter. In most of the cases, they proved to be better or comparable with the existing commercial filter materials, adsorbents, and conventional methods used for decontamination of drinking water. Being natural and their abundance presence makes them a low-cost green, nontoxic adsorbent which can be used for removal of different contaminants from water and making clean and pure drinking water available for developed and developing nations.

# FUTURE DIRECTIONS

The use of clay materials with natural polymer coating holds great promise for water treatment. As mentioned above, the adsorption capacity of natural and modified clay minerals increases with the coating of polymer on them. More research is needed to get abundant results in using the hybrid clay and polymeric materials in water treatment. Another research which needs immediate attention involves using clay for successful removal of emerging contaminants present in trace amount in our drinking water. Present conventional water treatment technologies are incapable of removing the emerging contaminants. Currently, the research in this field is scarce. But the available research results hold significant promise for the use of modified clay materials for emerging contaminant treatment without undesired toxic effects to the ecosystem.

# ACKNOWLEDGMENTS

The author is grateful to Ms. Melanie Magre for her review comments on the initial draft of the paper. The author also wants to thank the reviewer for useful comments to improve the paper. The author is grateful to Professor Thomas J. Gerik for providing financial support for the publication of this paper.

# REFERENCES

1. The Clay Mineral Group, 2011,http://mineral.galleries.com/ minerals/silicate/clays.htm.

2. S. H. Lin and R. S. Juang, "Heavy metal removal from water by sorption using surfactant-modified montmorillonite," Journal of Hazardous Materials B, vol. 92, no. 3, pp. 315–326, 2002. · ·

3. B. S. Krishna, D. S. R. Murty, and B. S. Jai Prakash, "Thermodynamics of chromium(VI) anionic species sorption onto surfactant-

modified montmorillonite clay," Journal of Colloid and Interface Science, vol. 229, no. 1, pp. 230–236, 2000. ··

4.　S. E. Bailey, T. J. Olin, R. M. Bricka, and D. D. Adrian, "A review of potentially low-cost sorbents for heavy metals," Water Research, vol. 33, no. 11, pp. 2469–2479, 1999. ··

5.　S. Babel and T. A. Kurniawan, "Low-cost adsorbents for heavy metals uptake from contaminated water: a review," Journal of Hazardous Materials B, vol. 97, no. 1–3, pp. 219–243, 2003.

6.　R. L. Virta, U.S. Geological Survey-Minerals Information, 1996, http://minerals.usgs.gov/minerals/pubs/commodity/clays/190496.pdf.

7.　T. J. Pinnavaia, "Intercalated clay catalysts," Science, vol. 220, no. 4595, pp. 365–371, 1983.

8.　F. Cadena, R. Rizvi, and R. W. Peters, "Feasibility studies for the removal of heavy metal from solution using tailored bentonite, hazardous and industrial wastes," in Proceedings of the 22nd Mid-Atlantic Industrial Waste Conference, pp. 77–94, Drexel University, July 1990.

9.　K. Tanabe, "Solid acid and base catalysis," in Catalysis—Science and Technology, J. R. Anderson and M. Boudart, Eds., p. 231, Springer, New York, NY, USA, 1981.

10.　H. van Olphen, An Introduction to Clay Colloid Chemistry, Wiley Interscience, New York, NY, USA, 2nd edition, 1977.

11.　K. G. Bhattacharyya and S. Sen Gupta, "Adsorption of a few heavy metals on natural and modified kaolinite and montmorillonite: a review," Advances in Colloid and Interface Science, vol. 140, no. 2, pp. 114–131, 2008. ···

12.　W. Wang, A. Li, J. Zhang, and A. Wang, "Study on superabsorbent composite: XI. Effect of thermal treatment and acid activation of attapulgite on water absorbency of poly(acrylic acid)/attapulgite superabsorbent composite," Polymer Composites, vol. 28, no. 3, pp. 397–404, 2007. ··

13.　A. Li, J. Zhang, and A. Wang, "Preparation and slow-release property of a poly(acrylic acid)/attapulgite/sodium humate superabsorbent composite," Journal of Applied Polymer Science, vol. 103, no. 1, pp. 37–45, 2007. ··

14.    S. Ekici, Y. I ıkver, and D. Saraydın, "Poly (acrylamide-sepiolite) composite hydrogels: preparation, swelling and dye adsorption properties," Polymer Bulletin, vol. 57, no. 2, pp. 231–241, 2006. · ·

15.    E. M. Araújo, T. J. A. Mélo, L. N. L. Santana et al., "The influence of organo-bentonite clay on the processing and mechanical properties of nylon 6 and polystyrene composites," Materials Science and Engineering B, vol. 112, no. 2-3, pp. 175–178, 2004. · ·

16.    S. Pandey and S. B. Mishra, "Organic-inorganic hybrid of chitosan/ organoclay bionanocomposites for hexavalent chromium uptake," Journal of Colloid and Interface Science, vol. 361, no. 2, pp. 509–520, 2011. · ·

17.    G. J. Churchman, "Formation of complexes between bentonite and different cationic polyelectrolytes and their use as sorbents for non-ionic and anionic pollutants," Applied Clay Science, vol. 21, no. 3-4, pp. 177–189, 2002. · ·

18.    C. Breen, "The characterisation and use of polycation-exchanged bentonites," Applied Clay Science, vol. 15, no. 1-2, pp. 187–219, 1999. · ·

19.    A. Radian and Y. G. Mishael, "Characterizing and designing polycation—clay nanocomposites as a basis for imazapyr controlled release formulations," Environmental Science and Technology, vol. 42, no. 5, pp. 1511–1516, 2008. · ·

20.    D. Zadaka, S. Nir, A. Radian, and Y. G. Mishael, "Atrazine removal from water by polycation-clay composites: effect of dissolved organic matter and comparison to activated carbon," Water Research, vol. 43, no. 3, pp. 677–683, 2009. · · ·

21.    M. Darder, M. Colilla, and E. Ruiz-Hitzky, "Chitosan-clay nanocomposites: application as electrochemical sensors," Applied Clay Science, vol. 28, no. 1–4, pp. 199–208, 2005. · ·

22.    M. Darder, M. López-Blanco, P. Aranda, A. J. Aznar, J. Bravo, and E. Ruiz-Hitzky, "Microfibrous chitosan—sepiolite nanocomposites,"Chemistry of Materials, vol. 18, no. 6, pp. 1602–1610, 2006. · ·

23.    E. Ruiz-Hitzky, M. Darder, and P. Aranda, "Functional biopolymer nanocomposites based on layered solids," Journal of Materials Chemistry, vol. 15, no. 35-36, pp. 3650–3662, 2005. · ·

24.  J. H. An and S. Dultz, "Adsorption of tannic acid on chitosan-montmorillonite as a function of pH and surface charge properties,"Applied Clay Science, vol. 36, no. 4, pp. 256–264, 2007. · ·

25.  J. M. Li, X. G. Meng, C. W. Hu, and J. Du, "Adsorption of phenol, p-chlorophenol and p-nitrophenol onto functional chitosan," Bioresource Technology, vol. 100, no. 3, pp. 1168–1173, 2009. · · ·

26.  S. Sen Gupta and K. G. Bhattacharyya, "Removal of Cd(II) from aqueous solution by kaolinite, montmorillonite and their poly(oxo zirconium) and tetrabutylammonium derivatives," Journal of Hazardous Materials, vol. 128, no. 2-3, pp. 247–257, 2006. · · ·

27.  M. Ulmanu, E. Marañón, Y. Fernández, L. Castrillón, I. Anger, and D. Dumitriu, "Removal of copper and cadmium ions from diluted aqueous solutions by low cost and waste material adsorbents," Water, Air, and Soil Pollution, vol. 142, no. 1–4, pp. 357–373, 2003. ·

28.  K. G. Bhattacharyya and S. Sen Gupta, "Adsorption of chromium(VI) from water by clays," Industrial and Engineering Chemistry Research, vol. 45, no. 21, pp. 7232–7240, 2006. · ·

29.  O. Yavuz, Y. Altunkaynak, and F. Guzel, "Removal of copper, nickel, cobalt and manganese from aqueous solution by kaolinite," Water Research, vol. 37, no. 4, pp. 948–952, 2003. · ·

30.  K. G. Bhattacharyya and S. Sen Gupta, "Adsorption of Co(II) from aqueous medium on natural and acid activated kaolinite and montmorillonite," Separation Science and Technology, vol. 42, no. 15, pp. 3391–3418, 2007. · ·

31.  S. H. Lin and R. S. Juang, "Heavy metal removal from water by sorption using surfactant-modified montmorillonite," Journal of Hazardous Materials B, vol. 92, no. 3, pp. 315–326, 2002. · ·

32.  K. G. Bhattacharyya and S. Sen Gupta, "Influence of acid activation on adsorption of Ni(II) and Cu(II) on kaolinite and montmorillonite: kinetic and thermodynamic study," Chemical Engineering Journal, vol. 136, no. 1, pp. 1–13, 2008. · ·

33.  K. G. Bhattacharyya and S. Sen Gupta, "Adsorption of Fe(III) from water by natural and acid activated clays: studies on equilibrium

isotherm, kinetics and thermodynamics of interactions," Adsorption, vol. 12, no. 3, pp. 185–204, 2006. · ·

34. S. Sen Gupta and K. G. Bhattacharyya, "Interaction of metal ions with clays: I. A case study with Pb(II)," Applied Clay Science, vol. 30, no. 3-4, pp. 199–206, 2005. · ·

35. K. G. Bhattacharyya and S. Sen Gupta, "Pb(II) uptake by kaolinite and montmorillonite in aqueous medium: influence of acid activation of the clays," Colloids and Surfaces A, vol. 277, no. 1–3, pp. 191–200, 2006. · ·

36. S. Sen Gupta and K. G. Bhattacharyya, "Adsorption of Ni(II) on clays,"Journal of Colloid and Interface Science, vol. 295, no. 1, pp. 21–32, 2006. · · ·

37. A. Mellah and S. Chegrouche, "The removal of zinc from aqueous solutions by natural bentonite," Water Research, vol. 31, no. 3, pp. 621–629, 1997. · ·

38. L. C. A. Oliveira, R. V. R. A. Rios, J. D. Fabris, K. Sapag, V. K. Garg, and R. M. Lago, "Clay-iron oxide magnetic composites for the adsorption of contaminants in water," Applied Clay Science, vol. 22, no. 4, pp. 169–177, 2003. ·

39. Ö. Etci, N. Bekta , and M. S. Öncel, "Single and binary adsorption of lead and cadmium ions from aqueous solution using the clay mineral beidellite," Environmental Earth Sciences, vol. 61, no. 2, pp. 231–240, 2010. · ·

40. H. Gecol, P. Miakatsindila, E. Ergican, and R. H. Sage, "Biopolymer coated clay particles for the adsorption of tungsten from water,"Desalination, vol. 197, no. 1–3, pp. 165–178, 2006. · ·

41. S. Aytas, M. Yurtlu, and R. Donat, "Adsorption characteristic of U(VI) ion onto thermally activated bentonite," Journal of Hazardous Materials, vol. 172, no. 2-3, pp. 667–674, 2009. · · ·

42. P. C. Mishra and R. K. Patel, "Removal of lead and zinc ions from water by low cost adsorbents," Journal of Hazardous Materials, vol. 168, no. 1, pp. 319–325, 2009. · · ·

43. P. Yuan, M. Fan, D. Yang et al., "Montmorillonite-supported magnetite nanoparticles for the removal of hexavalent chromium [Cr(VI)] from aqueous solutions," Journal of Hazardous Materials, vol. 166, no. 2-3, pp. 821–829, 2009. · · ·

44. Y. S. Kang, S. Risbud, J. F. Rabolt, and P. Stroeve, "Synthesis and characterization of nanometer-size $Fe_3O_4$ and $-Fe_2O_3$ particles,"Chemistry of Materials, vol. 8, no. 9, pp. 2209–2211, 1996.

45. M. J. Angove, B. B. Johnson, and J. D. Wells, "The influence of temperature on the adsorption of cadmium(II) and cobalt(II) on kaolinite," Journal of Colloid and Interface Science, vol. 204, no. 1, pp. 93–103, 1998. · · ·

46. B. Doušová, L. Fuitová, T. Grygar et al., "Modified aluminosilicates as low-cost sorbents of As(III) from anoxic groundwater," Journal of Hazardous Materials, vol. 165, no. 1–3, pp. 134–140, 2009. · · ·

47. J. Fang, B. Deng, and T. M. Whitworth, "Arsenic removal from drinking water using clay membranes," ACS Symposium Series, vol. 915, pp. 294–305, 2006.

48. S. M. I. Sajidu, I. Persson, W. R. L. Masamba, E. M. T. Henry, and D. Kayambazinthu, "Removal of $Cd^{2+}$, $Cr^{3+}$, $Cu^{2+}$, $Hg^{2+}$, $Pb^{2+}$ and $Zn^{2+}$cations and $AsO_4^{3-}$ anions from aqueous solutions by mixed clay from Tundulu in Malawi and characterisation of the clay," Water SA, vol. 32, no. 4, pp. 519–526, 2006.

49. D. Peak, U. K. Saha, and P. M. Huang, "Selenite adsorption mechanisms on pure and coated montmorillonite: an EXAFS and XANES spectroscopic study," Soil Science Society of America Journal, vol. 70, no. 1, pp. 192–203, 2006. · ·

50. N. Zhang, L. S. Lin, and D. Gang, "Adsorptive selenite removal from water using iron-coated GAC adsorbents," Water Research, vol. 42, no. 14, pp. 3809–3816, 2008. · · ·

51. D. G. Barceloux, "Selenium," Journal of Toxicology—Clinical Toxicology, vol. 37, no. 2, pp. 145–172, 1999. · ·

52. N. Bleiman and Y. G. Mishael, "Selenium removal from drinking water by adsorption to chitosan-clay composites and oxides: Batch and columns tests," Journal of Hazardous Materials, vol. 183, no. 1–3, pp. 590–595, 2010. · · ·

53. P. Na, X. Jia, B. Yuan et al., "Arsenic adsorption on Ti-pillared montmorillonite," Journal of Chemical Technology and Biotechnology, vol. 85, no. 5, pp. 708–714, 2010. · ·

54.  C. L. Ake, K. Mayura, H. Huebner, G. R. Bratton, and T. D. Phillips, "Development of porous clay-based composites for the sorption of lead from water," Journal of Toxicology and Environmental Health A, vol. 63, no. 6, pp. 459–475, 2001. · ··

55.  G. Zhao, H. Zhang, Q. Fan et al., "Sorption of copper(II) onto super-adsorbent of bentonite-polyacrylamide composites," Journal of Hazardous Materials, vol. 173, no. 1–3, pp. 661–668, 2010. · ··

56.  Y. Bulut, G. Akçay, D. Elma, and I. E. Serhatlı, "Synthesis of clay-based superabsorbent composite and its sorption capability," Journal of Hazardous Materials, vol. 171, no. 1–3, pp. 717–723, 2009. · ··

57.  D. Thakre, S. Rayalu, R. Kawade, S. Meshram, J. Subrt, and N. Labhsetwar, "Magnesium incorporated bentonite clay for defluoridation of drinking water," Journal of Hazardous Materials, vol. 180, no. 1–3, pp. 122–130, 2010. · ·

58.  S. P. Kamble, P. Dixit, S. S. Rayalu, and N. K. Labhsetwar, "Defluoridation of drinking water using chemically modified bentonite clay," Desalination, vol. 249, no. 2, pp. 687–693, 2009. ·

59.  Y. X. Ma, F. M. Shi, X. L. Zheng, J. Ma, and J. M. Yuan, "Defluoridation from aqueous solutions by Zr-loaded bentonite," Journal of Harbin Institute of Technology (New Series), vol. 12, supplement 1, pp. 224–229, 2005.

60.  C. J. Mena-Duran, M. R. Sun Kou, T. Lopez et al., "Nitrate removal using natural clays modified by acid thermoactivation," Applied Surface Science, vol. 253, no. 13, pp. 5762–5766, 2007. ·

61.  L. Gu, X. Yu, J. Xu, L. Lv, and Q. Wang, "Removal of dichloroacetic acid from drinking water by using adsorptive ozonation," Ecotoxicology, vol. 20, no. 5, pp. 1160–1166, 2011. · ·

62.  J. Lu and J. Pan, "Removal of carbon tetrachloride from contaminated groundwater environment by adsorption method," in Proceedings of the 4th International Conference on Bioinformatics and Biomedical Engineering (iCBBE ‹10), Chengdu, China, June 2010. ·

63.  S. M. Rivera-Jimenez, M. M. Lehner, W. A. Cabrera-Lafaurie, and A. J. Hernández-Maldonado, "Removal of naproxen, salicylic

acid, clofibric acid, and carbamazepine by water phase adsorption onto inorganic-organic-intercalated bentonites modified with transition metal cations,"Environmental Engineering Science, vol. 28, no. 3, pp. 171–182, 2011. ·

64. H. B. Senturk, D. Ozdes, A. Gundogdu, C. Duran, and M. Soylak, "Removal of phenol from aqueous solutions by adsorption onto organomodified Tirebolu bentonite: equilibrium, kinetic and thermodynamic study," Journal of Hazardous Materials, vol. 172, no. 1, pp. 353–362, 2009. · ·

65. L. Gu, X. Zhang, L. Lei, and X. Liu, "Concurrent removal of humic acid and o-dichlorobenzene in drinking water by combined ozonation and bentonite coagulation process," Water Science and Technology, vol. 60, no. 12, pp. 3061–3068, 2009. · ·

66. J. Q. Jiang and C. G. Kim, "Comparison of algal removal by coagulation with clays and Al-based coagulants," Separation Science and Technology, vol. 43, no. 7, pp. 1677–1686, 2008. ·

67. Z. Gao, X. Peng, H. Zhang, Z. Luan, and B. Fan, "Montmorillonite-Cu(II)/Fe(III) oxides magnetic material for removal of cyanobacterial Microcystis aeruginosa and its regeneration," Desalination, vol. 247, no. 1–3, pp. 337–345, 2009. ·

68. D. Zadaka, S. Nir, A. Radian, and Y. G. Mishael, "Atrazine removal from water by polycation-clay composites: effect of dissolved organic matter and comparison to activated carbon," Water Research, vol. 43, no. 3, pp. 677–683, 2009. · ·

69. T. Undabeytia, S. Nir, T. Sánchez-Verdejo, J. Villaverde, C. Maqueda, and E. Morillo, "A clay-vesicle system for water purification from organic pollutants," Water Research, vol. 42, no. 4-5, pp. 1211–1219, 2008. · ·

70. G. Rytwo, Y. Kohavi, I. Botnick, and Y. Gonen, "Use of CV- and TPP-montmorillonite for the removal of priority pollutants from water,"Applied Clay Science, vol. 36, no. 1–3, pp. 182–190, 2007. ·

71. F. P. Bonina, M. L. Giannossi, L. Medici, C. Puglia, V. Summa, and F. Tateo, "Adsorption of salicylic acid on bentonite and kaolin and release experiments," Applied Clay Science, vol. 36, no. 1–3, pp. 77–85, 2007. ·

72. T. Wang, R. L. Zhu, F. Ge, J. X. Zhu, H. P. He, and W. X. Chen, "Sorption of phenol and nitrobenzene in water by CTMAB/CPAM organobentonites," Huanjing Kexue/Environmental Science, vol. 31, no. 2, pp. 385–389, 2010.

73. W. Zhang, Y. Ding, S. A. Boyd, B. J. Teppen, and H. Li, "Sorption and desorption of carbamazepine from water by smectite clays,"Chemosphere, vol. 81, no. 7, pp. 954–960, 2010. · ·

74. R. J. Morris, D. E. Williams, H. A. Luu, C. F. B. Holmes, R. J. Andersen, and S. E. Calvert, "The adsorption of microcystin-LR by natural clay particles," Toxicon, vol. 38, no. 2, pp. 303–308, 2000. ·

75. A. K. Chaturvedi, K. P. Yadava, K. C. Pathak, and V. N. Singh, "Defluoridation of water by adsorption on fly ash," Water, Air, and Soil Pollution, vol. 49, no. 1-2, pp. 41–69, 1990.

76. M. G. Sujana, R. S. Thakur, and S. B. Rao, "Removal of fluoride from aqueous solution by using alum sludge," Journal of Colloid and Interface Science, vol. 206, no. 1, pp. 94–101, 1998. · ·

77. A. Toyoda and T. Taira, "A new method for treating fluorine wastewater to reduce sludge and running costs," IEEE Transactions on Semiconductor Manufacturing, vol. 13, no. 3, pp. 305–309, 2000.

78. S. Ayoob and A. K. Gupta, "Fluoride in drinking water: a review on the status and stress effects," Critical Reviews in Environmental Science and Technology, vol. 36, no. 6, pp. 433–487, 2006. ·

79. WHO (World Health Organization), Fluorine and Fluorides, Environmental Health Criteria, Geneva, Switzerland, World Health Organization, 1984.

80. H. H. Murray, "Traditional and new applications for kaolin, smectite, and palygorskite: a general overview," Applied Clay Science, vol. 17, no. 5-6, pp. 207–221, 2000. ·

81. M. Sanchez Camazano and M. J. Sanchez Martin, "Factors influencing interactions of organophosphorus pesticides with montmorillonite,"Geoderma, vol. 29, no. 2, pp. 107–118, 1983. ·

82. C. C. Ainsworth, J. M. Zachara, and R. L. Schmidt, "Quinoline sorption on Na-montmorillonite: contributions of the protonated and neutral species," Clays & Clay Minerals, vol. 35, no. 2, pp. 121–128, 1987.

83. J. M. Rodriguez, A. J. Lopez, and S. Bruque, "Interaction of phenamiphos with montmorillonite," Clays & Clay Minerals, vol. 36, no. 3, pp. 284–288, 1988.

84. H. T. Shu, D. Li, A. A. Scala, and Y. H. Ma, "Adsorption of small organic pollutants from aqueous streams by aluminosilicate-based microporous materials," Separation and Purification Technology, vol. 11, no. 1, pp. 27–36, 1997. ·

85. A. Torrents and S. Jayasundera, "The sorption of nonionic pesticides onto clays and the influence of natural organic carbon," Chemosphere, vol. 35, no. 7, pp. 1549–1565, 1997. ·

86. T. G. Danis, T. A. Albanis, D. E. Petrakis, and P. J. Pomonis, "Removal of chlorinated phenols from aqueous solutions by adsorption on alumina pillared clays and mesoporous alumina aluminum phosphates,"Water Research, vol. 32, no. 2, pp. 295–302, 1998. ·

87. I. K. Konstantinou, T. A. Albanis, D. E. Petrakis, and P. J. Pomonis, "Removal of herbicides from aqueous solutions by adsorption on Al-pillared clays, Fe-Al pillared clays and mesoporous alumina aluminum phosphates," Water Research, vol. 34, no. 12, pp. 3123–3136, 2000. ·

88. D. Sun, W. Cai, C. Shi, X. Mu, Y. Song, and H. Qi, "Advanced oxidations of chloroacetic acids present in drinking water," Journal of Environmental Science and Health A, vol. 35, no. 10, pp. 1811–1816, 2000.

89. M. G. Pervova, V. E. Kirichenko, and K. I. Pashkevich, "Determination of chloroacetic acids in drinking water by reaction gas chromatography," Journal of Analytical Chemistry, vol. 57, no. 4, pp. 326–330, 2002. ·

90. World Health Organization, Guideline for Safe Recreational Water Environments 1, Coastal and Fresh Water, WHO, Geneva, Switzerland, 2003.

91. Health Canada, Summary of Guideline for Canadian Drinking Water Quality, Federal Provincial Territorial Committee on Drinking Water, Ottawa, Canada, 2003.

92. American Public Health Association/American Water Works Association/Water Environmental Federation, Standard Methods for the Examination of Water and Wastewater, Washington, DC, USA, 20th edition, 1998.

93.  T. Kiely, D. Donaldson, and A. Grube, Pesticide Industry Sales and Usage: 2000 and 2001 Market Estimates, U.S. Environmental Protection Agency, Washington, DC, USA, 2004.

94.  N. Graziano, M. J. Mcguire, A. Roberson, C. Adams, H. Jiang, and N. Blute, "2004 national atrazine occurrence monitoring program using the abraxis ELISA method," Environmental Science and Technology, vol. 40, no. 4, pp. 1163–1171, 2006. ·

95.  M. J. Cerejeira, P. Viana, S. Batista et al., "Pesticides in Portuguese surface and ground waters," Water Research, vol. 37, no. 5, pp. 1055–1063, 2003. ·

96.  A. Papastergiou and E. Papadopoulou-Mourkidou, "Occurrence and spatial and temporal distribution of pesticide residues in groundwater of major corn-growing areas of Greece (1996-1997)," Environmental Science and Technology, vol. 35, no. 1, pp. 63–69, 2001. ·

97.  J. M. S. van Maanen, M. A. J. de Vaan, A. W. F. Veldstra, and W. P. A. M. Hendrix, "Pesticides and nitrate in groundwater and rainwater in the Province of Limburg in the Netherlands," Environmental Monitoring and Assessment, vol. 72, no. 1, pp. 95–114, 2001. ·

98.  J. B. Sass and A. Colangelo, "European Union bans atrazine, while the United States negotiates continued use," International Journal of Occupational and Environmental Health, vol. 12, no. 3, pp. 260–267, 2006.

99.  J. Lemi , D. Kova evi , M. Tomaševi - anovi , D. Kova evi , T. Stani , and R. Pfend, "Removal of atrazine, lindane and diazinone from water by organo-zeolites," Water Research, vol. 40, no. 5, pp. 1079–1085, 2006. · ·

100.  M. J. Sanchez-Martin, M. S. Rodriguez-Cruz, M. S. Andrades, and M. Sanchez-Camazano, "Efficiency of different clay minerals modified with a cationic surfactant in the adsorption of pesticides: influence of clay type and pesticide hydrophobicity," Applied Clay Science, vol. 31, no. 3-4, pp. 216–228, 2006. ·

101.  M. Borisover, E. R. Graber, F. Bercovich, and Z. Gerstl, "Suitability of dye-clay complexes for removal of non-ionic organic compounds from aqueous solutions," Chemosphere, vol. 44, no. 5, pp. 1033–1040, 2001. ·

102. G. Abate and J. C. Masini, "Adsorption of atrazine, hydroxyatrazine, deethylatrazine, and deisopropylatrazine onto Fe(III) polyhydroxy cations intercalated vermiculite and montmorillonite," Journal of Agricultural and Food Chemistry, vol. 53, no. 5, pp. 1612–1619, 2005. · ·

103. M. Streat and L. A. Sweetland, "Removal of pesticides from water using hyper cross linked polymer phases: part 2-Sorption studies," Process Safety and Environmental Protection, vol. 76, pp. 127–134, 1998.

104. G. J. Churchman, "Formation of complexes between bentonite and different cationic polyelectrolytes and their use as sorbents for non-ionic and anionic pollutants," Applied Clay Science, vol. 21, no. 3-4, pp. 177–189, 2002. ·

105. C. Breen, "The characterisation and use of polycation-exchanged bentonites," Applied Clay Science, vol. 15, no. 1-2, pp. 187–219, 1999. ·

106. A. Radian and Y. G. Mishael, "Characterizing and designing polycation—clay nanocomposites as a basis for imazapyr controlled release formulations," Environmental Science and Technology, vol. 42, no. 5, pp. 1511–1516, 2008. ·

107. W. W. Carmichael, "Freshwater cyanobacteria (blue-green algal) toxins," in Natural Toxins: Characterization, Pharmacology and Therapeutics, C. L. Ownby and G. V. Odell, Eds., pp. 3–16, Pergamon Press, London, UK, 1988.

108. P. Cohen and P. T. W. Cohen, "Protein phosphatases come of age," Journal of Biological Chemistry, vol. 264, no. 36, pp. 21435–21438, 1989.

109. S. Yoshizawa, R. Matsushima, M. F. Watanabe et al., "Inhibition of protein phosphatases by microcystis and nodularin associated with hepatotoxicity," Journal of Cancer Research and Clinical Oncology, vol. 116, no. 6, pp. 609–614, 1990.

110. R. E. Honkanen, J. Zwiller, R. E. Moore et al., "Characterization of microcystin-LR, a potent inhibitor of type 1 and type 2A protein phosphatases," Journal of Biological Chemistry, vol. 265, no. 32, pp. 19401–19404, 1990.

111. C. MacKintosh, K. A. Beattie, S. Klumpp, P. Cohen, and G. A. Codd, "Cyanobacterial microcystin-LR is a potent and specific

inhibitor of protein phosphatases 1 and 2A from both mammals and higher plants,"FEBS Letters, vol. 264, no. 2, pp. 187–192, 1990. ·

112. R. Nishiwaki-Matsushima, S. Nishiwaki, T. Ohta et al., "Structure-function relationships of microcystins, liver tumor promoters, in interaction with protein phosphatase," Japanese Journal of Cancer Research, vol. 82, no. 9, pp. 993–996, 1991.

113. R. Nishiwaki-Matsushima, T. Ohta, S. Nishiwaki et al., "Liver tumor promotion by the cyanobacterial cyclic peptide toxin microcystin-LR,"Journal of Cancer Research and Clinical Oncology, vol. 118, no. 6, pp. 420–424, 1992. ·

114. H. Fujiki and M. Suganuma, "Tumor promotion by inhibitors of protein phosphatases 1 and 2A: the okadaic acid class of compounds," Advances in Cancer Research, vol. 61, pp. 143–194, 1993.

115. L. A. Lawton, B. J. P. A. Cornish, and A. W. R. MacDonald, "Removal of cyanobacterial toxins (microcystins) and cyanobacterial cells from drinking water using domestic water filters," Water Research, vol. 32, no. 3, pp. 633–638, 1998. ·

116. K. I. Harada and K. Tsuji, "Persistence and decomposition of hepatotoxic microcystins produced by cyanobacteria in natural environment," Journal of Toxicology—Toxin Reviews, vol. 17, no. 3, pp. 385–403, 1998.

Chapter

# 4

# The Role of Fillers on Friction and Slide Wear Characteristics in Glass-epoxy Composite Systems

B. Suresha[1], G. Chandramohan[1], J. N. Prakash[2], V. Balusamy[3], and K.Sankaranarayanasamy[4]

[1]Department of Mechanical Engineering, PSG College of Technology, Coimbatore-641 004, INDIA

[2]Research and Development Centre, East Point College of Engineering. and Technology, Bangalore-560 025, INDIA

[3]Department of Metallurgical Engineering, PSG College of Technology, Coimbatore-641 004, INDIA

[4]Department of Mechanical Engineering, National Institute of Technology, Trichy-620 015, INDIA

# ABSTRACT

The comparative performance of Glass-Epoxy (G-E) composite systems interfaced with graded fillers has been examined. In this study, composite materials were experimentally investigated under varying load and sliding velocities by using a Pin-on-Disc type wear tester. The influence of two inorganic fillers, silicon carbide particles (SiC) and graphite, on the wear of the glass fabric reinforced epoxy composites under dry sliding conditions has been investigated. For increased load and sliding velocity situations, higher wear loss was recorded. Some of these observations are supplemented by scanning electron microscopic (SEM) investigations. The coefficients of frictional values show an increasing trend with subsequent increase in load/sliding velocities. It was observed that the Graphite filled G-E composite shows lower coefficient of friction than the other two composites irrespective of variation in the load/sliding velocities. SiC filled G-E composite exhibited the maximum wear resistance. Further, wear of the matrix, breakage of reinforcing fibers, matrix debris formation and interface separation were observed in unfilled and graphite-filled G-E composites. Other interesting SEM features have been noticed and discussed.

# INTRODUCTION

In recent times, there has been a remarkable growth in the large-scale production of fiber and/or filler reinforced epoxy matrix composites. Because of their high strength-toweight and stiffness-to-weight ratios, they are extensively used for a wide variety of structural applications as in aerospace, automotive and chemical industries [1]. On account of their good combination of properties, fiber reinforced polymer composites (FRPCs) are used for producing a number of mechanical components such as gears, cams, wheels, brakes, clutches, bearings and seals. Most of these are subjected to tribological loading conditions. The FRPCs exhibit relatively low densities and they can also be tailored for our design requirements by altering the stacking sequences to provide high strength and stiffness in the direction of high loading [2].

A number of material-processing strategies have been used to improve the wear performance of polymers. Glass fiber reinforced polymeric composites traditionally show poor wear resistance and high friction due to the brittle nature of the reinforcing fibers. This has prompted many researchers to cast the polymers with fibers/fillers. Considerable efforts are being made to extend the range of applications. Such use would provide economical and functional benefits to both manufacturers and consumers. Various researchers have studied the tribological behaviour of FRPCs. Studies have been conducted with various shapes, sizes, types and compositions of fibers in a number of matrices [3-8]. In general these materials exhibit lower wear and friction when compared to pure polymers. An understanding of the friction and wear mechanisms of FRPC's would aid in the development of a new class of materials so as to counter the challenges faced by researchers. Reviews of such works found in articles [9-11] have shown that the friction and wear behaviour of FRPCs exhibits anisotropic characteristics.

Use of inorganic fillers dispersed in polymeric composites is increasing. Fillers not only reduce the cost of the composites, but also meet performance requirements, which could not have been achieved by using reinforcement and resin ingredients alone. In order to obtain perfect friction and wear properties many researchers modified polymers using different fillers [12-20]. Briscoe et al. [12] reported that the wear rate of highdensity polyethylene (HDPE) was reduced with the addition of inorganic fillers, such as CuO and $Pb_3O_4$. Tanaka [13] concluded that the wear rate of polytetrofluroethylene (PTFE) was reduced when filled with $ZrO_2$ and $TiO_2$. Bahadur et al. [14-16] found that the compounds of copper such as CuO and CuS were very effective in reducing the wear rate of PEEK, PTFE, Nylon and HDPE. Kishore et al. [17] studied the influence of sliding velocity and load on the friction and wear behaviour of G-E composite, filled with either rubber or oxide particles, and reported that the wear loss increased with increase in load/speed. Solid lubricants such as graphite and $MoS_2$ [18, 19] when added to polymers proved to be effective in reducing the coefficient of friction and wear rate of composites.

The use of graphite as a particulate filler has been reported to improve tribological behavior in metal matrix composites (MMCs) [20].

Most of the above findings are based on either randomly oriented or unidirectionally oriented fiber composites. Woven fabric reinforced composites [21] are gaining popularity because of their balanced properties in the fabric plane as well as their ease of handling during fabrication. Mody et al. [22] have shown that the simultaneous existence of parallel and anti-parallel oriented carbon fibers in a woven configuration leads to a synergistic effect on the enhancement of the wear resistance of the composite.

The objective of this work is to investigate the friction and wear properties of particulate filled G-E composites sliding against a hardened steel counterface. As a comparison, the friction and wear properties of plain G-E were also evaluated under identical test conditions. This work helps in understanding the function of different fillers in G-E composites. This work is believed to be helpful for understanding the function of different fillers in G-E composites.

# EXPERIMENTAL

## Materials

Woven glass fabrics made of 360 gsm, containing E-glass fibers of diameter 5-10 μm has been employed. The matrix system used is a medium viscosity epoxy resin (LAPOX L-12) and a room temperature curing polyamine hardener (K-6) both supplied by ATUL India Ltd, Gujarat, India. The fillers that have been used are silicon carbide (SiC) and graphite particulates.

## Fabrication

All laminates used in this study were manufactured by dry hand layup technique. E-glass plain weave roving fabric, which is compatible to epoxy resin, is used as the reinforcement. The epoxy resin is mixed with the hardener in the ratio 100:12 by weight. The stacking procedure consists of placing the fabric one above the other with the resin mix well spread between the fabrics. A porous teflon film is placed on the completed stack. To ensure uniform thickness of the sample a spacer of size 3 mm is used. The mold plates have a release agent smeared on it. The whole assembly is pressed in a hydraulic press (0.5 MPa) and allowed to cure for a day at room temperature. After demolding, post curing was done at 120°C for 2 h using an electrical oven. The laminate so prepared has a size 250 mm X 250 mm X 3 mm. To prepare the filled G-E composites, filler (SiC and Graphite) is mixed with a known amount of epoxy resin. The details of the composites are shown in Table 1. The test samples are cut to size 5 mm x 5 mm x 3 mm with the help of a diamond tipped cutter.

**Table 1**: Details of samples prepared

| Sample code | Matrix | Reinforcement | Filler | wt. % |
|---|---|---|---|---|
| A | Epoxy | E-glass fabric | Graphite | 5 |
| B | Epoxy | E-glass fabric | SiCp | 5 |
| C | Epoxy | E-glass fabric | ---- | ---- |

# Test Procedure

A pin-on-disc test setup was used for slide wear experiments. The surface of the sample (5 mm X 5 mm) glued to a pin of dimensions 6 mm diameter and 22 mm length comes in contact with a hardened disc of hardness 62 $HR_C$. The counter surface disc was made of En 32 steel having dimensions of 165 mm diameter, 8 mm thick and surface roughness (Ra) of 0.84 µm. The test was conducted on a track of 115 mm diameter for a specified test duration, load and velocity [23]. Prior to testing, the test samples were rubbed against a 600-grade SiC paper. The surfaces of both the sample and the disc were cleaned with a soft paper soaked in acetone before the test. The pin assembly was initially weighed using a digital electronic balance (0.1 mg accuracy). The test was carried out by applying normal load (30 N to 70 N) and run for a constant sliding distance (5000 m) at different sliding velocities (3, 4 and 5 m/s). At the end of the test, the pin assembly was again weighed in the same balance. The difference between the initial and final weights was a measure of slide wear loss. A minimum of three trials was conducted to ensure repeatability of test data. The friction force at the sliding interface of the specimen was measured at an interval of 5 minutes using a frictional load cell. The coefficient of friction was obtained by dividing the frictional force by the applied normal force. Selected samples were coated with a thin layer of gold on the worn surface and subjected to microscopic examination using scanning electron microscope.

# RESULTS AND DISCUSSION

Experimental data on the slide wear loss of filled and unfilled G-E composite samples are shown in Figs. 1 to 3 for different loads (30 to 70 N) and sliding velocities (3 to 5 m/s). Table 2 shows the results pertaining to the coefficient of friction of filled and unfilled G-E composite system. It is observed from the figures and Table 2 that there is a strong inter-dependence between the friction coefficients and wear loss irrespective of the loads and sliding velocities employed. The SEM photographs of select combinations of filled and unfilled G-E samples subjected to slide wear are shown in Figs. 4, 5 and 6 respectively.

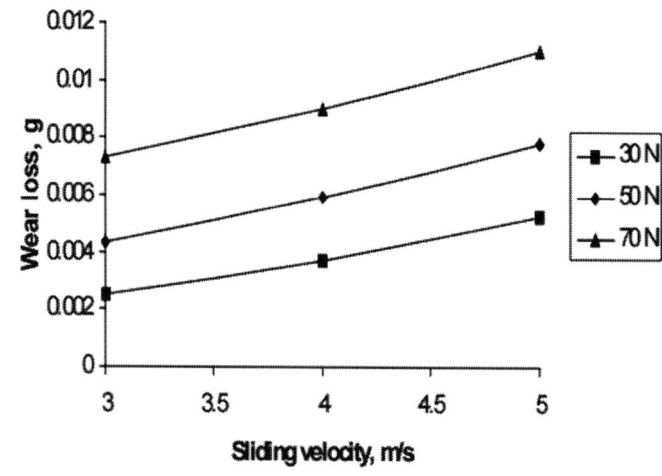

**Figure 1:** Wear loss versus sliding velocity of "A" type sample.

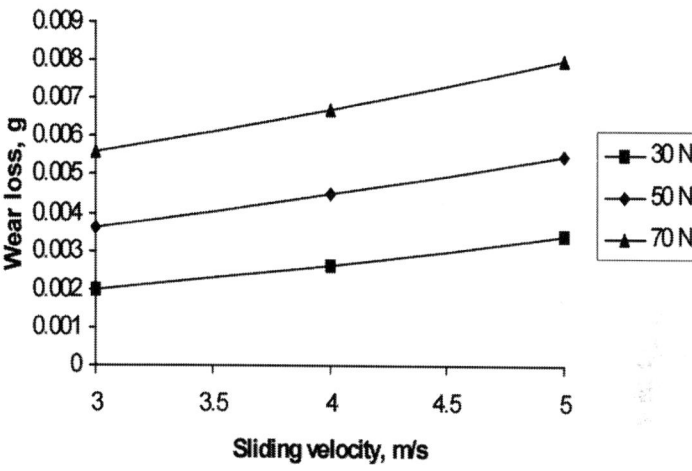

**Figure 2:** Wear loss versus sliding velocity of "B" type sample.

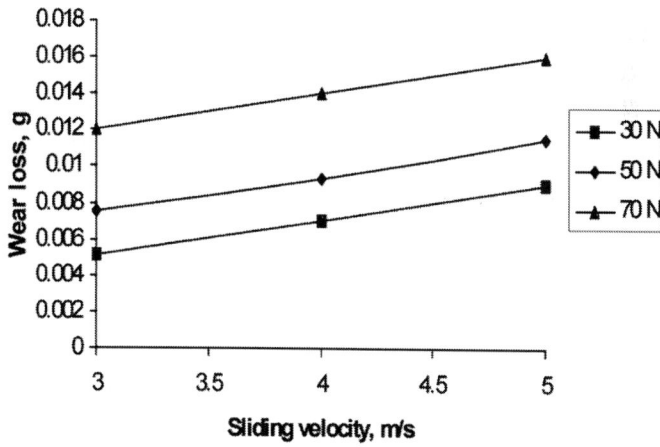

**Figure 3**: Wear loss versus sliding velocity of "C" type sample.

**Table 2:** Coefficient of friction of samples tested

| Load, | Coefficient of friction | | | | | | | | |
|---|---|---|---|---|---|---|---|---|---|
| N | Sliding velocity | | | Sliding velocity | | | Sliding velocity | | |
| | 3 m/s | | | 4 m/s | | | 5 m/s | | |
| | A | B | C | A | B | C | A | B | C |
| 30 N | 0.30 | 0.37 | 0.38 | 0.31 | 0.38 | 0.41 | 0.33 | 0.40 | 0.43 |
| 50 N | 0.33 | 0.40 | 0.42 | 0.38 | 0.45 | 0.47 | 0.37 | 0.51 | 0.52 |
| 70 N | 0.35 | 0.50 | 0.51 | 0.45 | 0.52 | 0.54 | 0.42 | 0.55 | 0.61 |

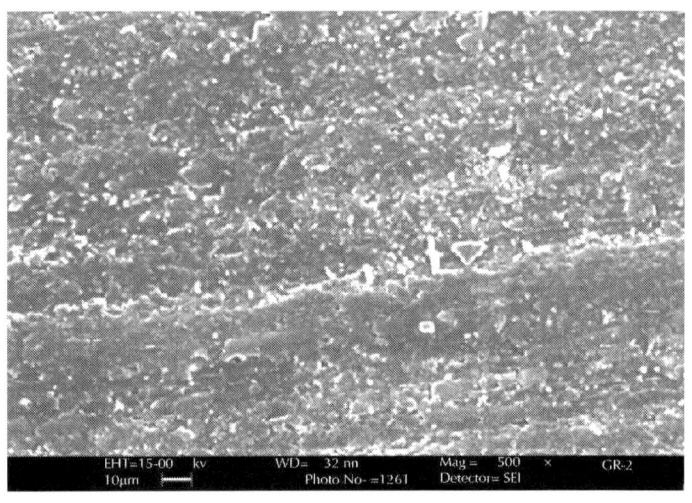

(a)

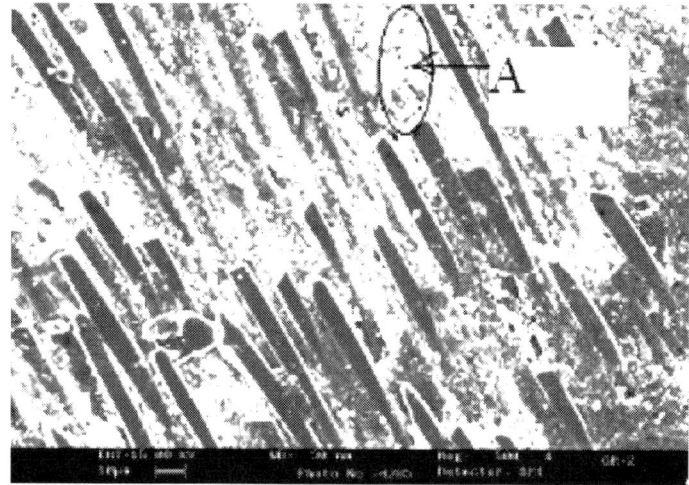

(b)

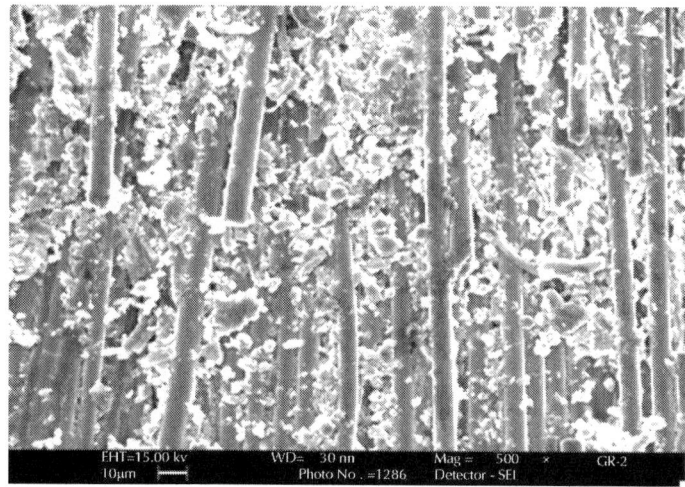

(c)

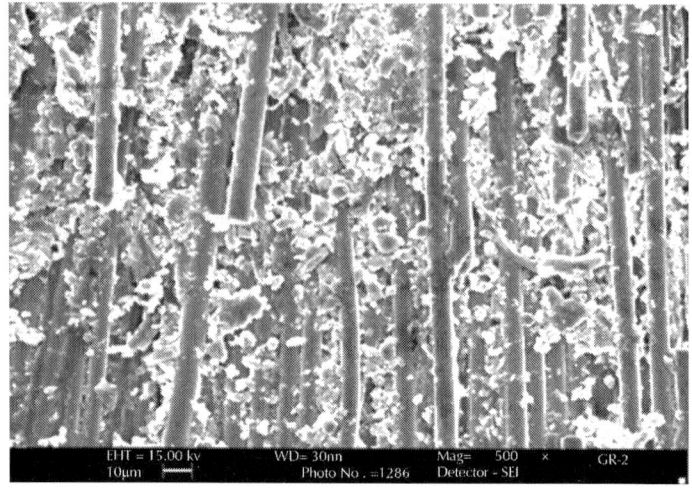

(d)

**Figure 4**: SEM picture of "A" sample at: (a) 30 N, 3 m/s, (b) 30 N, 5 m/s. (c) 70N, 3 m/s, and (d) 70 N, 5 m/s.

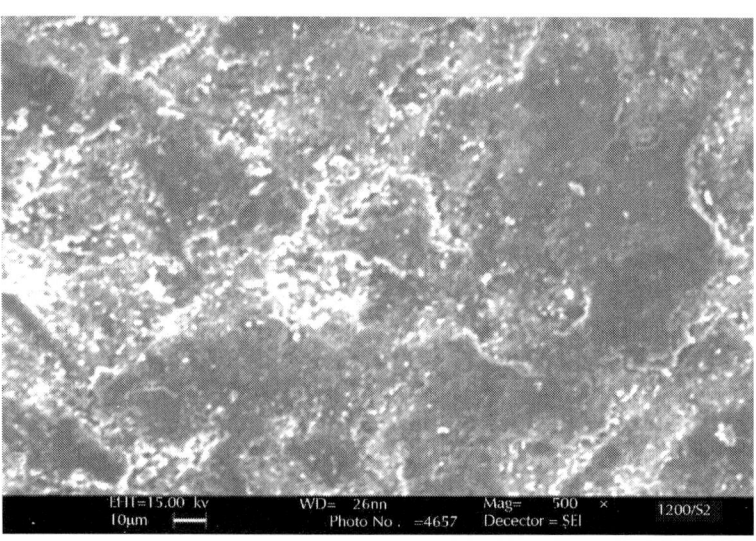

(a)

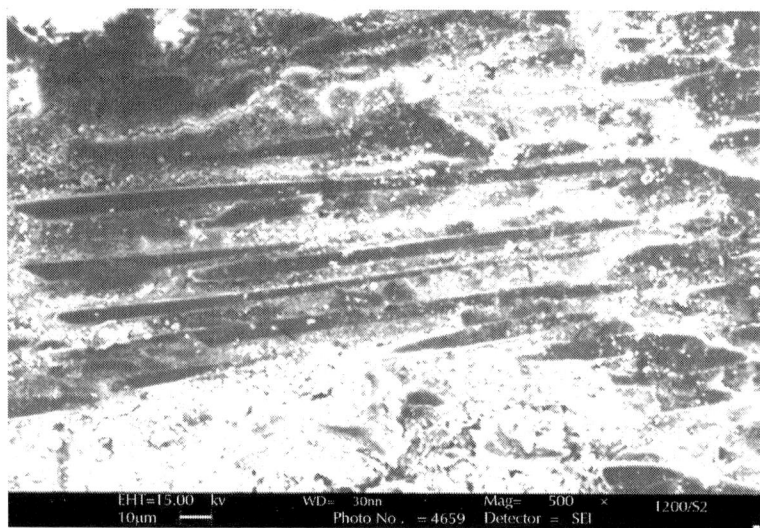

(b)

(c)

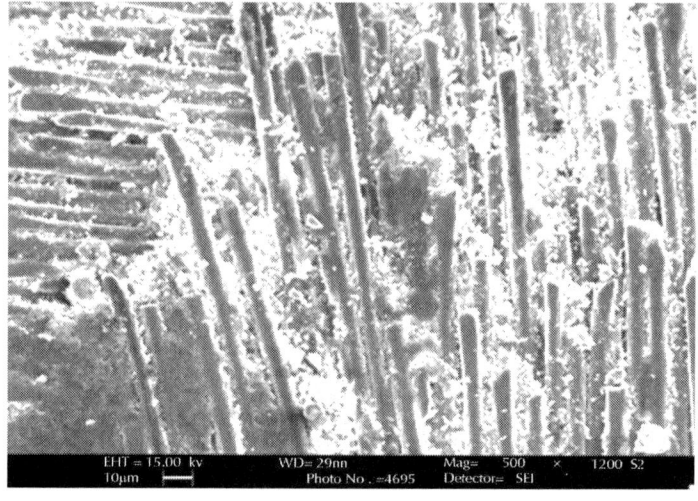

(d)

**Figure 5:** SEM picture of "B" sample at: (a) 30 N, 3 m/s, (b) 30 N, 5 m/s. (c) 70N, 3 m/s, and (d) 70 N, 5 m/s.

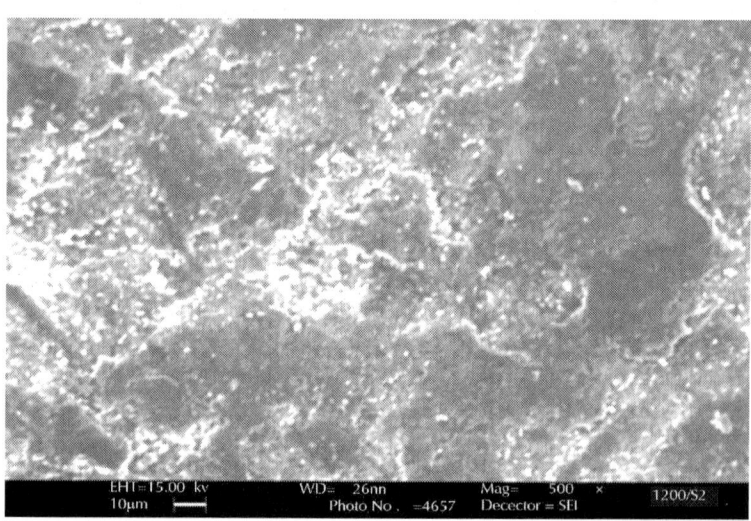

(a)

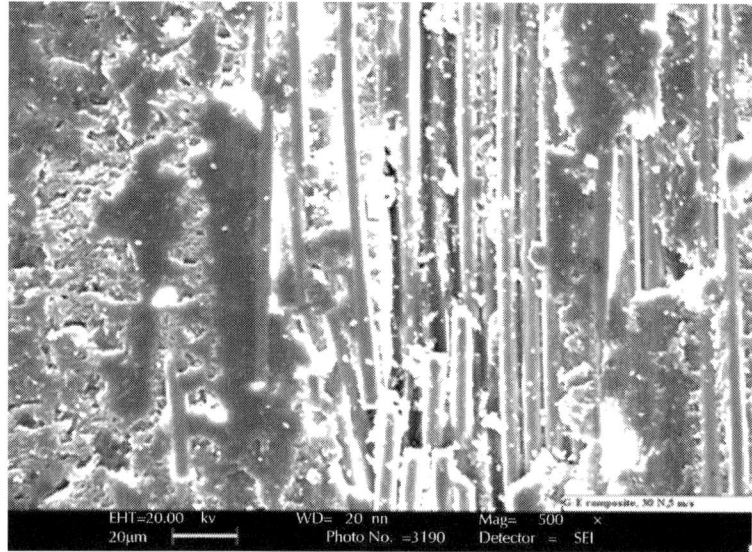

(b)

(c)

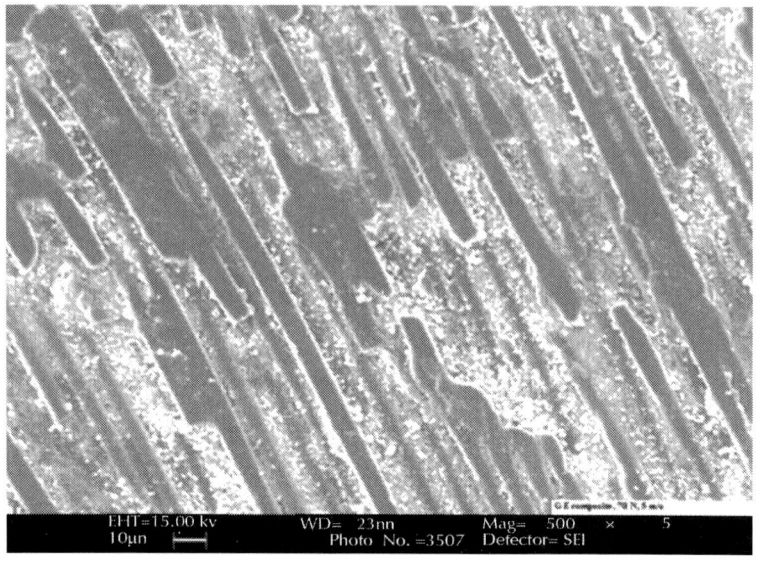

(d)

**Figure 6:** SEM picture of "C" sample at: (a) 30 N, 3 m/s, (b) 30 N, 5 m/s. (c) 70N, 3 m/s, and (d) 70 N, 5 m/s.

# Coefficient of Friction

The variation in coefficient of friction with varying sliding velocities/ loads of filled and unfilled G-E composites is shown in Table 2. For the filled G-E composites, an increasing trend in the coefficient of friction is seen, with increase in sliding velocity/load. However, comparison of graphite-filled G-E (sample A) with SiC-G-E (sample B) and G-E composites (sample C) indicate that the coefficient of friction of graphite filled G-E composite (sample A) is less. The reduction in coefficient of friction is attributed to the presence of graphite particulates acting as a solid lubricant. On the other hand, the increase in coefficient of friction in SiC-G-E may occur because of the inclusion of hard silicon carbide particles.

For the unfilled G-E samples, like in filled G-E samples, an increase in sliding velocity/load results in increase in the coefficient of friction. The increase in coefficient of friction is due to the fact that easy detachment of softened epoxy from the reinforcement and more breakage of reinforced glass fibers.

## Slide Wear Data

The slide wear data of filled and unfilled G-E composites shown in Figs. 1 to 3 are considered for interpretation. The results reveal that the wear loss increases with increase in sliding velocity irrespective of the load employed both for filled and unfilled G-E composite systems. However, the magnitudes of wear loss values are much less in filled G-E samples compared to unfilled G-E samples at all loads.

It is seen from Figs. 1 to 3 that the filler in G-E composites appears to influence the friction and wear behaviour. The wear losses of the composites decreases with filler addition and show the maximum wear resistance (least wear loss) for SiC-filled G-E and the least for unfilled G-E).

For C-type sample, the fibers are oriented parallel to the sliding surface and also to the sliding direction. In this position the fibers can be easily detached from the matrix; hence it is observed that the increase in wear loss is much greater than that observed in filled G-E composite samples. Further, increased exposure of the reinforcement of fibers to the counter surface results in increased fiber fracture due to the frictional thrust.

## Scanning Electron Microscopy

The slide wear data in respect of select samples are discussed based on the scanning electron microscopic features. The SEM pictures of Sample A shown in Figs. 4(a) and 4(b) pertaining to the test conditions of 30 N, 3 m/s and 30 N, 5 m/s respectively are considered for interpretation. Thus Fig. 4(a) shows less wear of the matrix, matrix covering the debris and hardly any breakage of fibers. Fig. 4(b) on the other hand, shows higher degree of fiber breakage, mostly cleavage type of smaller size and smearing of debris on the fibers (marked 'A' in Fig. 4(b)). The observations corroborate the wear data reported in Fig. 1. For the same sample at higher load of 70 N, keeping the sliding velocity at 3 m/s and 5 m/s respectively, Fig. 4c and 4d show the corresponding SEM features.

Thus, Fig. 4(c) displays increased debris formation, agglomeration of debris, adherence of debris on the fibers and more number of breakage of fibers compared to the condition seen in Fig. 4(a) (30 N, 3 m/s). Figure 4(d) records increased breakage of fibers with cleavage type of fracture, interface separation between the fibers and matrix and debris formation concentrated at specific locations. Further, few fibers along the sliding direction are found to be disoriented. There is one to one correspondence between the SEM observations and wear test results.

The SEM picture of B-type sample subjected to a load of 30 N and sliding velocity 3 m/s is shown in Fig. 5(a). The spread of the matrix and fewer wear debris formation are noticed. Both delamination and debonding increased with increasing sliding velocity (5 m/s), which resulted in exposure of reinforced fibers (Fig. 5(b)) along the sliding direction. In Fig. 5(c), it is noticed that the debris begins to cluster around the fibers. When Fig. 5(d) is compared with the Fig. 5(b), it is obvious that with application of higher load and no change in sliding velocity there has been observed increase of more breakage of fibers and further the broken fibers show inclined type of fracture. These observations are in accordance with experimental wear test data presented in Fig. 2.

The SEM features of C- type samples (plain G-E composite) subjected to various loads and sliding velocities are displayed in Figs. 6(a) to 6(d). Fig. 6(a) shows the SEM picture of the sample subjected to 30 N load and a sliding velocity of 3 m/s. It is observed that the matrix wear is more and fiber exposure is less (compare Fig. 6(a) with Fig. 4(a) and 5(a)). These features support the wear behaviour as seen in Fig. 3. The SEM features pertaining to 30 N load and 5 m/s sliding velocity is shown in Fig. 6(b), which reveals that the matrix debris is well spread, yielding more number of glass fiber breakages compared to the sample subjected to 30 N load and 3 m/s sliding velocity (Fig. 6(a)). By increasing the load from 30 N to 70 N and keeping the sliding velocity at 3 m/s, the wear surface features of sample C shows higher wear of the matrix, increased fiber-matrix debonding as well as breakage of fibers as shown in Fig. 6(c). Increasing the velocity to 5 m/s for the same sample (Fig. 6(d)) results in higher matrix debris formation and heavy breakage of glass fibers (cleavage type) in large numbers compared to Fig. 6(c) (70 N, 3 m/s). Also, the wear debris is getting totally distributed. These SEM photographs corroborate the wear data shown in Fig. 3.

# CONCLUSIONS

The following inferences are drawn from the above study.

- Inclusion of Graphite and SiC particulate fillers contributed significantly in reducing friction and exhibited better wear resistant properties.

- Silicon carbide filled G-E composite shows higher resistance to slide wear compared to plain G-E composites.

- There has been an observed marked improvement in wear resistance as seen in SiCG-E composite sample compared to plain G-E sample.

- Graphite filled G-E composite shows lower coefficient of friction compared to the other two samples. The reduction in coefficient of friction in A-type sample can be attributed to the presence of Graphite which acts as a solid lubricant.

- SEM examinations of worn surfaces show that the type of wear changed from adhesive wear to abrasive wear for all the samples tested.

- Increased wear resistance and reduced coefficient of friction are positive traits, which make the composite suitable to be used as liners in coal handling equipments.

# ACKNOWLEDGEMENTS

The authors are grateful to the Additional Director Dr. S. Seetharamu of Central Power Research Institute, Materials Testing Division, Bangalore for extending the laboratory facilities for the present study. The authors thank the Central Power Research Institute management for the permission extended to publish this paper.

# REFERENCES

1.    ASM Hand book, 1992, Materials Park, Ohio, USA, ASM International, Volume 18.

2.    Pascoe, M.W., 1973, "Plain and filled plastics materials in bearing: a review." Tribology, Vol. 6 No. 5, pp. 184-190.

3.   Sung, N.H., and Suh, N.P., 1979, "Effect of Fiber orientation on friction and wear of fiber reinforced polymeric composites." Wear, Vol. 53, pp. 129-141.

4.   Chang, H.W., 1983, "Wear characteristics of composite: effect of fiber orientation." Wear, Vol. 85, No. 1, pp. 81-91.

5.   Suresha, B., Chandramohan, G., Samapthkumaran, P., Seetharamu, S., and Vynatheya, S., 2006, "Friction and wear characteristics of carbon-epoxy and glass-epoxy woven roving fiber composites." Journal of Reinforced Polymers and composites, Vol. 25, pp. 771-782.

6.   Bijwe, J., Tewari, U.S., and Vasudevan, P., 1989 "Friction and wear studies of short glass fiber reinforced polythermide composite." Wear, Vol. 132, pp. 247-264.

7.   Viswanth, B., Verma, A.P., and Rao, C.V.S.K., 1991, "Effect of fiber geometry on friction and wear of glass fiber-reinforced composites." Wear, Vol. 145, pp. 315-327.

8.   Tripaty, B.S., Furey, M.J., 1993, "Tribological behaviour unidirectional graphiteepoxy and carbon -PEEK composites. Wear, Vol. 162-164, pp. 385-396.

9.   El-Sayed, A.A, El-Sherbiny, M.J., Abo-El-Ezz, A.S., Aggag, G.A., 1995, "Friction and wear properties of polymeric composite materials for bearing applications." Wear, Vol. 184, pp. 45-53.

10.  Cirino, M., Friedrich, K., and Pipes, R.B., 1988, "The effect of fiber orientation on the abrasive wear behaviour of polymer composite materials." Wear, Vol. 121, pp. 127-141.

11.  Lancaster, J. K., 1972, "Lubrication of carbon fiber-reinforced polymers : Part II— Organic fluids." Wear, Vol. 20, No. 3, pp. 335-351.

12.  Briscoe, B. J., Pogosion, A. K., and Tabor, D., 1974, "The friction and wear of high Density polyethylene; the action of lead oxide and copper oxide fillers." Wear, Vol. 27, pp. 19-34.

13.  Tanaka, K., 1986, Effect of various fillers on the friction and wear of PTFE-based composites, In: Friction and Wear of Polymer composites, Volume 205, pp. 137-174, (Friedrich K editor), Elsevier, Amsterdam.

14. Bahadur, S., Fu, Q., and Gong, D., 1994, "The effect of reinforcement and the synergism between CuS and carbon fiber on the wear of nylon." Wear, Vol. 178, pp. 123-130.

15. Bahadur, S., and Tabor, D., 1985, Role of fillers in friction and wear behaviour of HDPE In: Polymer wear and its control, Volume 287-268 (L.H. Lee (ed.) ACM symposium series, Washington DC.

16. Bahadur, S., Gong, D., Anderegg, J. W., 1992 "The role of copper composites as fillers in the transfer film formation and wear of Nylon." Wear, Vol. 154, pp. 207- 223.

17. Kishore, Sampathkumaran, P., Seetharamu, S., Vynatheya, S., Murali, A., Kumar, R. K., 2000, "SEM observations of the effect of velocity and load on the slide wear characteristics glass-fabric-epoxy composites with different fillers." Wear, Vol. 237, pp. 20-27.

18. Kishore, Sampathkumaran, P., Seetharamu. S., Thomas, P., Janardhana, M. A., 2005, "Study on the effect of the type and content of filler in epoxy-glass composite system on the friction and wear characteristics." Wear Vol. 259, pp. 634-641.

19. Wang, J., Gu, M., Songhao, Ge, S., 2003, "The role of the influence of MoS2 on the tribological properties of carbon fiber reinforced Nylon 1010 composites." Wear, Vol. 255, pp. 774-779.

20. Basavarajappa, S., Chandramohan, G. C., 2005, "Wear studies on metal matrix composites: A Taguchi Approach." J. of Materials Sci. and Tech., Vol. 21, No. 6, pp. 348-350.

21. Viswanath, B., Verma, A. P., and Kameswara Rao, C. V. S., 1992, "Effect of matrix content on strength and wear of woven roving glass polymeric composites." Comp Sci Tech., Vol. 44 pp. 77-86.

22. Mody, P. B., Chou, T. W., Friedrich, K., 1988, "Effect of testing conditions and microstructure on the sliding wear of graphite fiber/PEEK matrix composites." J. Mater. Sci., Vol. 23, pp. 4319-4330.

23. Annual hand book of ASTM standards, Section 3,03,02, ASTM G-99 (1995), Phildelphia, USA.

# A Comparative Study of the Phase Distribution in Carbon-Silica Hybrid Fillers for Rubber Obtained by Different Methods

Omar A. Al-Hartomy[1, 2], Ahmed A. Al-Ghamdi[1], Said A. Farha Al-Said[1, 2], Nikolay Dishovsky[3], Michael B. Ward[4], Petrunka Malinova[3], and Mihail Mihaylov[3]

[1]Department of Physics, Faculty of Science, King Abdulaziz University, Jeddah, Saudi Arabia

[2]Department of Physics, Faculty of Science, University of Tabuk, Tabuk, Saudi Arabia

[3]Department of Polymer Engineering, University of Chemical Technology and Metallurgy, Sofia, Bulgaria

[4]LEMAS, Institute for Materials Research, SPEME, University of Leeds, Leeds, UK

# ABSTRACT

Different types of carbon-silica fillers were prepared via pyrolysis-cum-water vapor of waste green tires tread and impregnation method. Dual phase fillers have been characterized by energy dispersive X-ray (EDX) spectroscopy in a scanning transmission electron microscope (STEM) or STEM-EDX. Phase distribution in hybrid fillers for rubber was investigated. The results achieved show that the conditions of obtaining influence the distribution and the location of the phases in the carbon-silica hybrid fillers as well as their most essential characteristics including specific area, oil absorption number, iodine adsorption number, ash content and others.

# INTRODUCTION

The primary filler factors influencing elastomer reinforcement are: 1) the primary particle size or specific surface area, which, together with loading, determines the effective contact area between the filler and polymer matrix; 2) the structure or the degree of irregularity of the filler unit, which plays an essential role in the restrictive motion of elastomer chains under strain; 3) the surface activity, which is the predominant factor with regard to filler-filler and filler-polymer interaction [1] . The effects of filler-filler and filler-elastomer interactions on rubber reinforcement are also described[1] [2] . In order to meet the constantly growing requirements of producers, particularly those of automobile tires, in respect of the proposed fillers, the company Cabot Corporation has presented a dual phase filler of the carbon-silica type, produced by aco-fuming process [3] -[5] . With this material, the filler-filler interaction is substantially reduced due to the surface modification, and the polymer-filler interaction is enhanced by increasing the surface energy of the carbon domain of the filler and creating chemical bonding via coupling reaction between polymer chains and silanols on the silica domain. In the last years we developed successfully two different alternative methods for obtaining of hybrid fillers of carbon-silica type: by pyrolysis-cum-water vapor of worn out, waste tires and by impregnation technology. Generally, the pyrolysis of tires is described as a method for their recycling [6] -[8] . We assumed that a dual phase filler will be obtained if not standard (filled with carbon black alone) tires are subjected to pyrolysis but the so called "green" tires, containing a big quantity of silica as a filler. We assumed also that carbon phase will be formed during the pyrolysis in a result of elastomer destruction. The method is attractive, most of all, for solving an essential environmental problem and at the same time dual phase filler is being obtained whose advantages have already been described above. The experiments, carried out by us, showed that our assumption had been correct. The obtained filler was characterized and its effect on elastomeric composites on various bases including those based on natural rubber, epoxidized natural rubber, ethylene-propylene triple copolymer and others [9] -[13] .

The impregnation method is one of the most widely used and popular methods for applying different phases on carriers (first of all in the production of catalysts) or for the purpose of surface and texture modification of adsorbents [14] . Using this method the second phase is deposited from the solution of so called "precursor" (the most often salts or soluble complexes) by impregnation. The impregnated substances are subjected to chemical or physical influence that may be considered like thermal activation. The method of impregnation applying of the second phase was chosen because it allowed controlled disposition of the second phase (silica) on the carbon black surface. A part of it may be retained also within the pores of the carbon black particles and in the spaces between the particles. In our previous investigations we obtained hybrid filler for rubber by deposition of zinc oxide phase on active carbon surface using the impregnation method [15] . In another previous paper [16] possibilities for application of impregnation method for obtaining of new type of reinforcing fillers for rubbers by introduction and deposition of silica phase on the particles of low active furnace carbon black, having low value of their specific surface were investigated, filler thus obtained was characterized and its influence on the properties of styrene butadiene rubber (SBR) based composites (applicable for passenger tread compounds) was studied. It is found that the mentioned above fillers increase considerably the modulus at 100% of elongation and tensile strength in comparison to standard carbon black containing composites.

One of the most important questions which interested us, was to compare what was the distribution of the phases of carbon and silica in the fillers obtained by described above different methods, to study their phase structure, to understand to what extent silica and carbon phases interpenetrated, how (and if) conditions of obtaining influenced the phase distribution and in what was the difference between these fillers. Finding answers to these questions is the purpose of present comparative study. Without a doubt, the most appropriate method for that is energy dispersive X-ray (EDX) spectroscopy in a scanning transmission electron microscope (STEM) or STEM-EDX. High angle annular dark field (HAADF) imaging in the STEM displays compositional contrast that results from different atomic number of the elements and their distribution. EDX allows one to identify what those particular elements are and their relative proportions. Initial EDX analysis usually involves the generation of an X-ray spectrum from the entire scan area of the STEM image. The Y-axis shows the counts (number of X-rays received and processed by the detector) and the X-axis shows the energy of those X-rays. In addition to this EDX software can:

-keep the electron beam stationary on a spot or series of spots and generate spectra that will provide more localized elemental information;

-have the electron beam follow a line drawn on the sample image and generate a plot of the relative proportions of different elements along that line (one dimensional line scanning). Modern software now collects entire cumulative spectra from each point, so element choices can be made post acquisition;

-map the distribution and relative proportion (intensity) of different elements over the scanned area (two dimensional mapping).

# EXPERIMENTAL

## Materials

### Carbon-Silica Fillers Preparation by Pyrolysis-cum-Water Vapor of Waste Green Tires [6] [7]

Two types of carbon-silica fillers (marked as CSF-1 and CSF-2 respectively) were obtained via pyrolysis-cumwater vapor of waste green tires tread (Michelin Energy 195/65 R15) in different conditions as follows:

CSF-1: temperature—500°C, water vapour concentration—40%;
CSF-2: temperature—700°C, water vapour concentration—40%.

## Carbon-Silica Fillers Preparation by Impregnation Technology

Two types of carbon-silica fillers (marked as CSF-10 and CSF-20 respectively) were obtained by impregnation procedure as follows:

Industrial furnace carbon black type PM-15 (produced in Russia, which characteristics are close to those of carbon black ASTM type N 776), with specific area of 15 $m^2/g$ were used as a substratum in our investigation. Silicasol (silica content—40%, pH—9 and density 1.3 $g/cm^3$) was chosen as impregnating agent due to our idea to obtain reinforcing carbon-silica dual phase filler. Impregnation modifying of carbon black with $SiO_2$ is accomplished by spraying with the help of an atomizer via Thin Layer Chromatography method under continuous stirring. The quantity of $SiO_2$ solution needed for sufficient wetting of carbon black is 1.6 mL/g carbon black. The quantity of silicasol for impregnation of carbon black is calculated preliminarily in order to be introduced the determined percentage of $SiO_2$. The procedure of impregnation treating of carbon black is proceeding in two stages. In the first stage of synthesis, one tenth of calculated solution is diluted with distilled water in a ratio 1:10. After spraying this solution on carbon black under permanent stirring, the sample remains for 24 hours at a room temperature, and then undergoes thermal treatment for 2 hours at a temperature of 323°K - 353°K and 2 hours at 523°K. Second stage: the rest of solution is applied by analogical procedure and conditions. After applying the solution, carbon black stays again for 24 hours in air at a room temperature, then being treated thermally for 2 hours at a temperature of 323°K - 353°K, four hours at 425°K and four hours at 523°K. The dual phase fillers obtained are of different $SiO_2$ content, calculated as follows: CSF-10—10% $SiO_2$, CSF-20—20% $SiO_2$.

# Methods Used for Characterization of Substratum and Carbon Silica Dual Phase Fillers

- Iodine adsorption—in correspondence to [17].
- Oil absorption number—in correspondence to [18].
- Specific surface area—calculated by the method of BET [19].

The ash content of the carbon silica fillers obtained by impregnation technology was determined according to [20] . The ash from the fillers was studied by complete silicate analyses, weight analysis, atomic absorption spectroscopy-AAS (Perkin Elmer 5000), inductively coupled plasma-optical emission spectroscopy-ICP-OES ("Prodigy" High Dispersion ICP-OES, Teledyne Lemas Labs).

For ascertaining the distribution of carbon black and silica, the hybrid products and their physical mixture were also investigated by the method of STEM-EDX (STEM, FEI Tecnai F20, FEGTEM) operating at 200 kV and equipped with an Energy Dispersive X-ray Analyzer (Oxford Instruments, 80 mm$^2$ X-max SDD, running INCA software). One gram of powder of each sample was used for the analyses. A small amount of it was mixed with solvent and then drop onto a carbon support film using a pipette. The ratio of the Si to C counts was used as an estimate of the relative ranking of the amount of silica and carbon in certain regions in the filler aggregates.

# RESULTS AND DISCUSSION

Table 1 and Table 2 summarize the main properties of the carbon-silica fillers obtained.

Table 3 and Table 4 summarize the results of silicate analysis, AAS and ICP-OES (in %) of ashes of the carbon-silica fillers investigated.

By comparing the results in Tables 1-4, following makes immediately an impression:

The results for the fillers obtained by the methods developed by us differ very strongly due to the radically opposing technologies used: by pyrolysis, initial raw material used was vulcanizate (worn automobile tires), containing significant amount of silica as filler due to which their ash content is very high. The content of organic matter (carbon black) is within the limits 28% - 36% and contents of inorganic-within 72% - 64%. The availability of some amounts of carbon black in the fillers investigated, allowed determining its iodine adsorption and oil numbers. The carbon phase occurring upon destruction of rubber in the tire in the conditions of pyrolysis-cum-water is characterized by very high values of iodine number. The iodine number reflects a "not true" surface area, because it is affected by porosity, surface impurities and surface oxidation (and oil number). Oil absorption number reflects the empty space (void volume) between the aggregates and agglomerates usually expressed as a volume of dibutylphtalate absorbed by a given amount of the filler (comparable with those of the most active furnace carbon black N220) which is indicative of high adsorption activity as regards rubber ma cromolecules, but also for a pronounced tendency to the formation of secondary structures-aggregates and agglomerates. The specific surface (BET nitrogen adsorption surface area shows the "total" surface area including porosity) is also high, similar to that of carbon black N220. The case with the dual phase fillers, obtained by impregnation is contrary: ash content is much lower, the inorganic matter is within the limits of 13% - 22%, i.e. organic matter prevails in them (within the limits of 78% - 87%), iodine and oil numbers have values specific for the lowest active furnaces carbon black, which is used as a raw material. For both types of fillers, silica prevails in the ash and its content has similar percentage values, but in the ashes of pyrolysis product there is a significant amount of zinc oxide (3% - 4%), entered in the rubber mixture due to its activating action in respect of vulcanization process, while its content in the ashes of impregnation product is minimal-under 0.01%. However, it is found in it higher levels of disodium oxide (2% - 2.3%), coming probably from silicasol, compared with pyrolysis product, where it is within the limits of 0.85% - 0.90%.

**Table 1:** Main properties of the carbon-silica fillers obtained by pyrolysis

| Properties | CSF-1 | CSF-2 |
|---|---|---|
| Iodine adsorption (IA), mg/g | 215 | 127 |
| Oil absorption number (OAN), mL/100g | 140 | 157 |
| Specific surface area (BET), m²/g | 116 | 108 |
| Ash content, % | 64 | 72 |

**Table 2**: Main properties of the carbon-silica fillers obtained by impregnation

| Properties | CSF-10 | CSF-20 |
|---|---|---|
| Iodine adsorption (IA), mg/g | 20 | 16 |
| Oil absorption number (OAN), mL/100g | 69 | 62 |
| Specific surface area (BET), m$^2$/g | 31 | 48 |
| Ash content, % | 13 | 22 |

**Table 3**: Silicate analysis, AAS and ICP-OES (in %) of the ashes of carbon-silica fillers obtained by pyrolysis

|  | CSF-1 | CSF-2 |
|---|---|---|
| $SiO_2$ | 92.56 | 93.84 |
| $Al_2O_3$ | 0.55 | 0.55 |
| CaO | 0.18 | 0.14 |
| MgO | 0.23 | 0.12 |
| $Na_2O$ | 0.86 | 0.91 |
| $K_2O$ | 0.08 | 0.05 |
| $Fe_2O_3$ | 0.46 | 0.40 |
| MnO | <0.01 | <0.01 |
| $TiO_2$ | 0.02 | 0.03 |
| ZnO | 3.82 | 2.8 |
| CuO | 0.08 | 0.08 |
| PbO | 0.03 | 0.02 |
| NiO | <0.01 | <0.01 |
| $Cr_2O_3$ | 0.01 | <0.01 |
| Heat losses, 1000°C | 1.07 | 0.83 |

**Table 4**: Silicate analysis, AAS and ICP-OES (in %) of ashes of the carbon-silica fillers obtained by impregnation

|  | CSF-10 | CSF-20 |
|---|---|---|
| $SiO_2$ | 94.73 | 95.26 |
| $Al_2O_3$ | 0.24 | 0.26 |
| CaO | 0.31 | 0.22 |
| MgO | 0.11 | 0.08 |
| $Na_2O$ | 2.30 | 1.97 |
| $K_2O$ | 0.35 | 0.19 |
| $Fe_2O_3$ | 0.11 | 0.13 |
| MnO | 0.07 | 0.01 |
| $TiO_2$ | <0.01 | 0.24 |
| ZnO | <0.01 | <0.01 |
| CuO | <0.01 | <0.01 |
| PbO | <0.01 | <0.01 |
| NiO | <0.01 | <0.01 |
| $Cr_2O_3$ | <0.01 | <0.01 |
| Heat losses, 1000°C | 1.72 | 1.61 |

Representative HAADF images and compositional maps of the carbon silica fillers obtained by pyrolysis and impregnation are shown in Figures 1-4.

The first general impression is that carbon and $SiO_2$ phases are much better mixed in the samples CSF-1 and CSF-2, prepared by the pyrolysis than in samples, prepared by impregnation. Regarding the actual type of mixing, it is quite clear that both phases exist individually, i.e. discreet regions of amorphous $SiO_2$, and of soot-like carbon. The only difference is in the degree to and scale on which the particles are mixed. The EDX maps show how the different particles were distributed. In the CSF-1 $SiO_2$ is more evenly distributed, the aggregates are smaller than and not as well-insulated as in CSF-2, the penetration of the phases one to another is more fully. It appears that there is better mixing in CSF-1 compared to CSF-2 from the maps and if this was the case it would agree with the results from the surface area measurements. One can see from the Table 1 that indeed the surface area of CSF-1 is higher than the surface area of CSF-2. The maps create the impression that the carbon phase occupies the spaces between the aggregates of silicon dioxide, mostly with spherical morphology, which is logical, since at the time of the formation of this phase through elastomer destruction, silica exists already as a phase. We consider this a positive effect, because in this way it is hindered the formation of large silica aggregates as a result of the strong interactions "filler-filler", which is not desired. In all samples investigated carbon and silica exist as separate phases and there are not evidences for some type of chemical mixing. One can see that in them carbon and silica coexist in space, penetrating one another. Carbon and silica are found together (in the same place) all over CSF-1 and CSF-2 samples, while that penetration in CSF-1 is at a higher level and the two phases are to be found oftener together at one and same time.

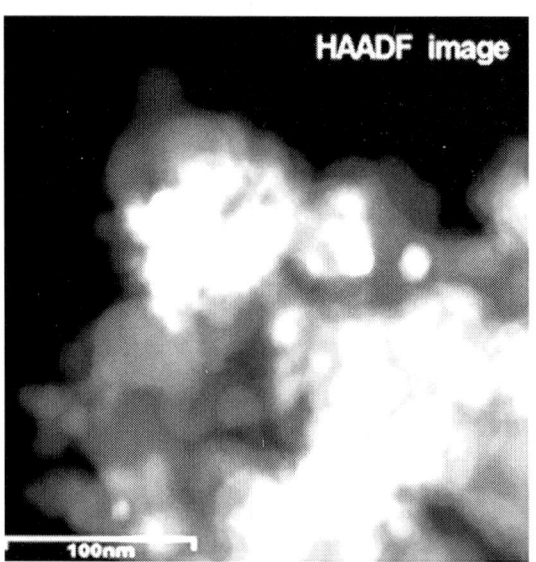

(a)

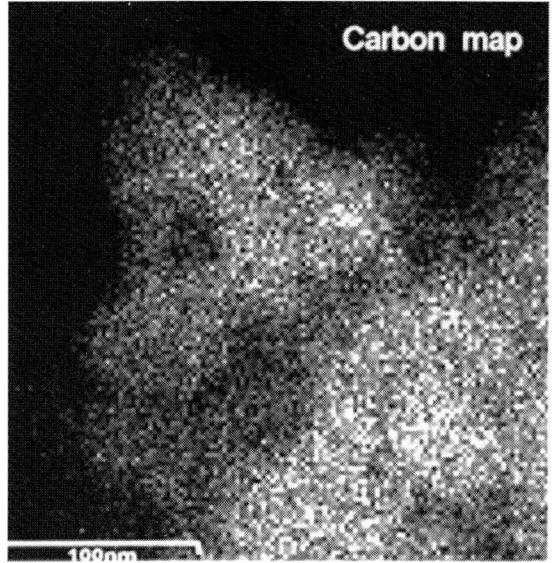

(b)

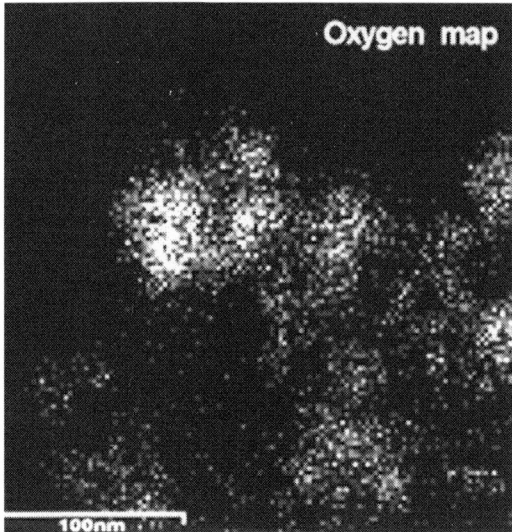

(c)

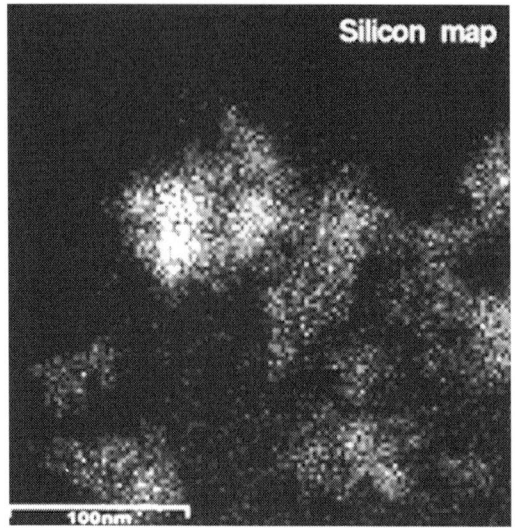

(d)

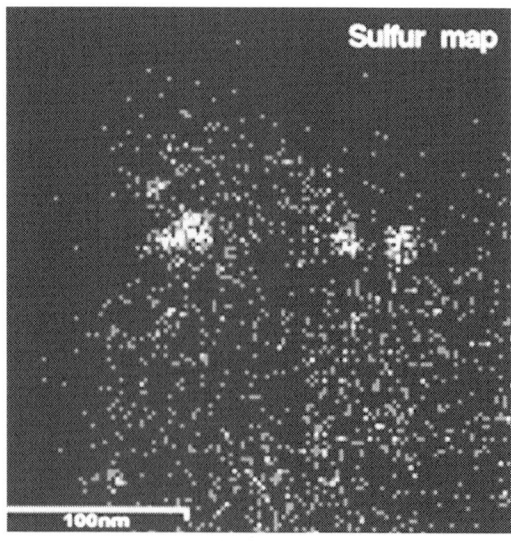

(e)

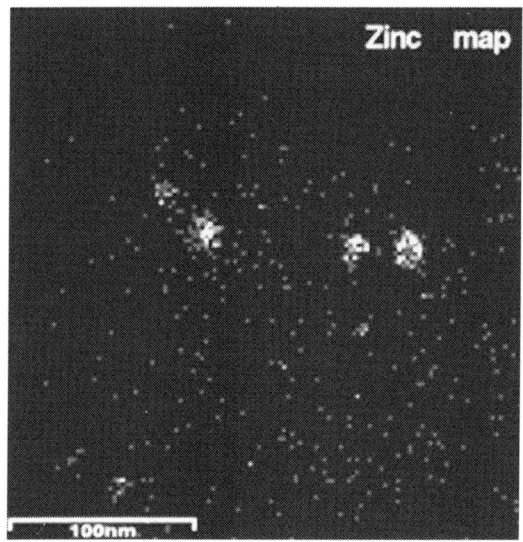

(f)

**Figure 1**: HAADF image (a) and compositional maps of CSF-1 (b), (c), (d), (e), (f).

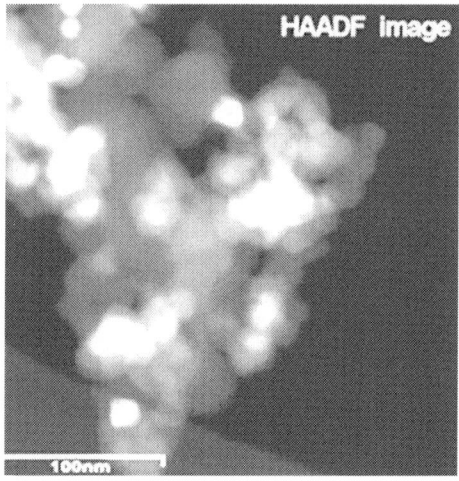

(a)

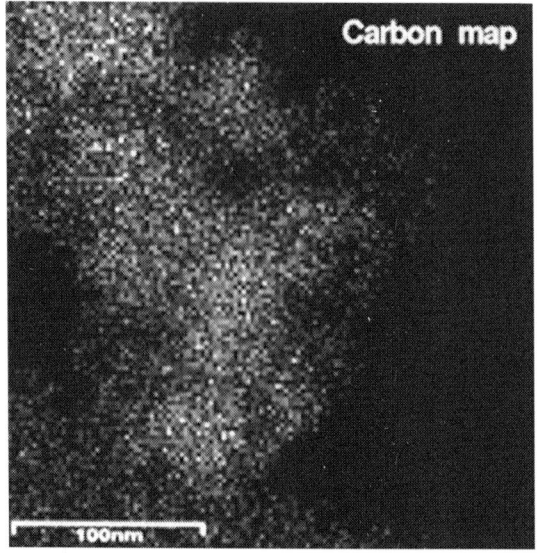

(b)

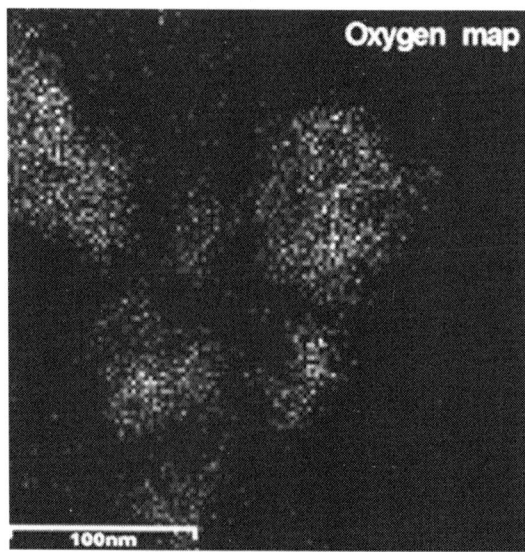

(c)

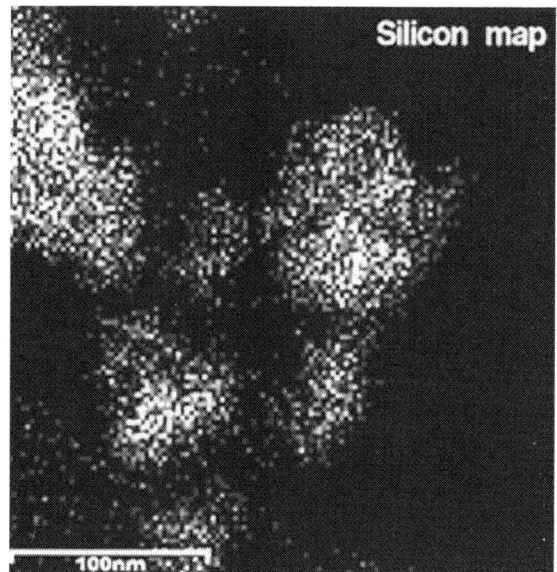

(d)

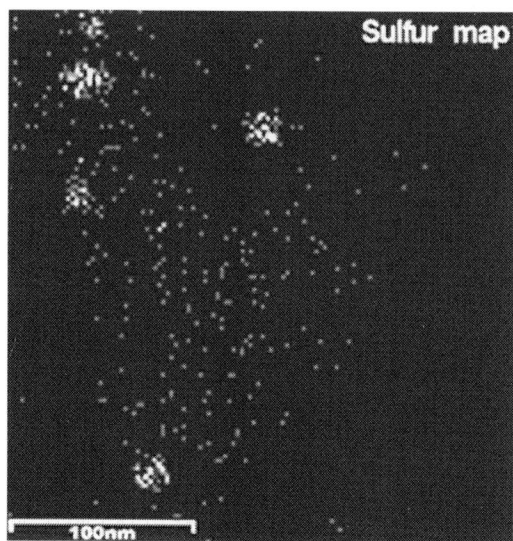

(e)

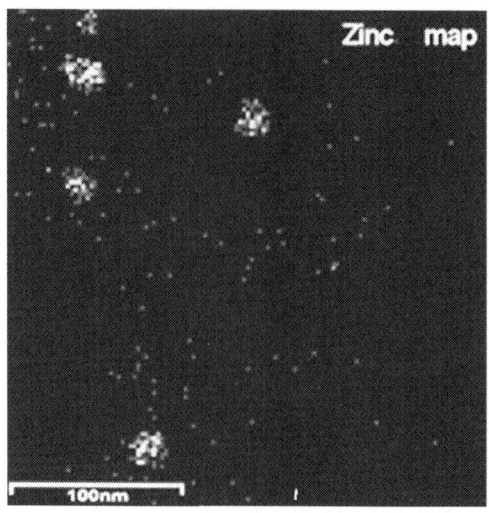

(f)

**Figure 2**: HAADF image (a) and compositional maps of CSF-2 (b), (c), (d), (e), (f).

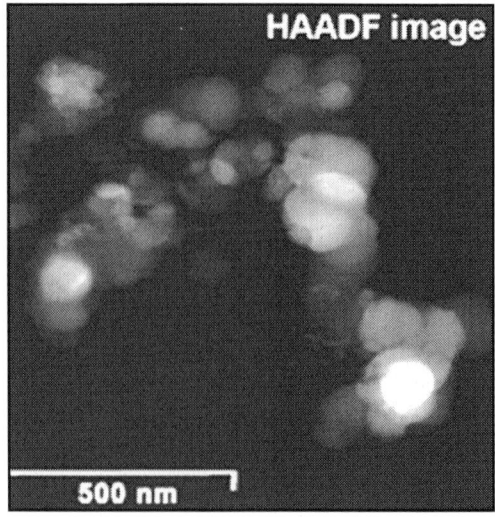

(a)

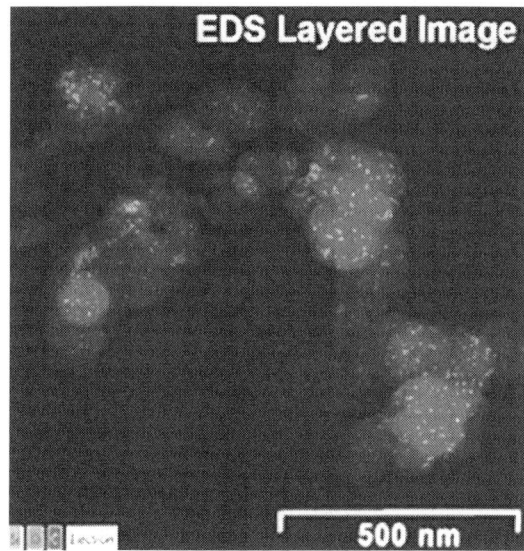

(b)

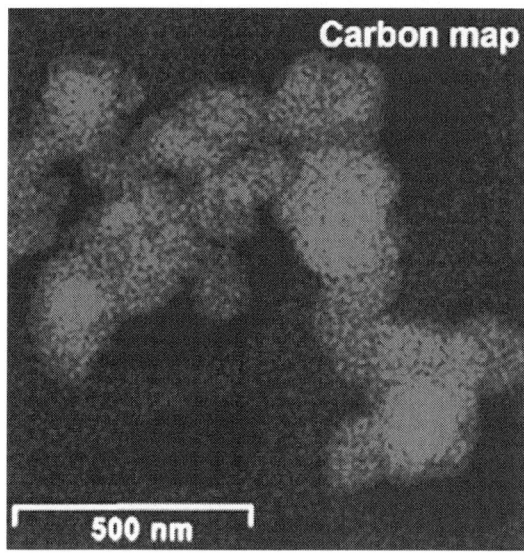

(c)

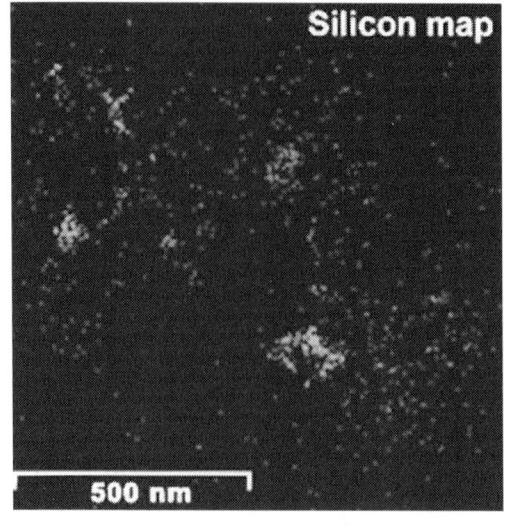

(d)

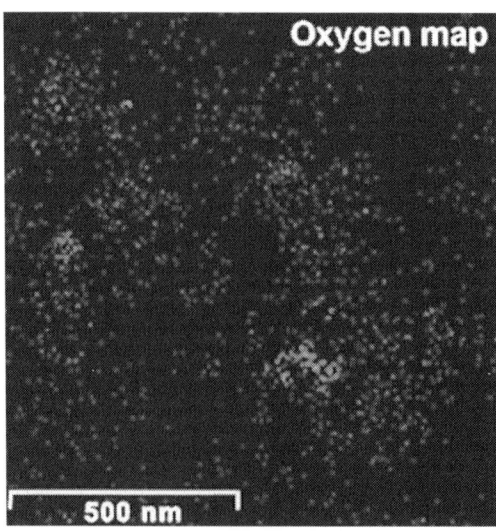

(e)

**Figure 3**: HAADF image and compositional maps of CSF-10.

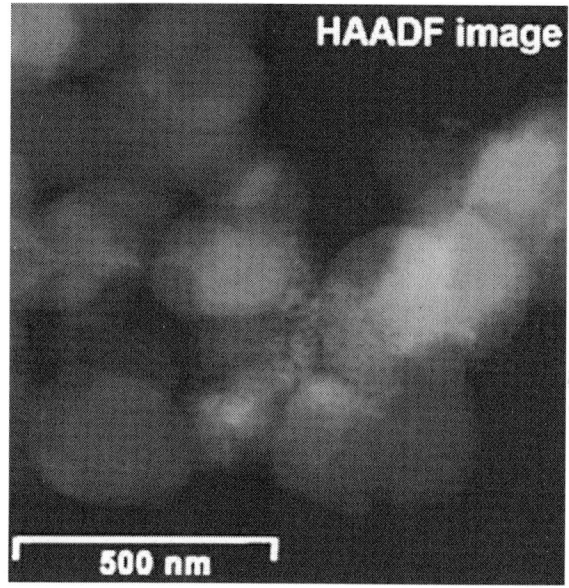

(a)

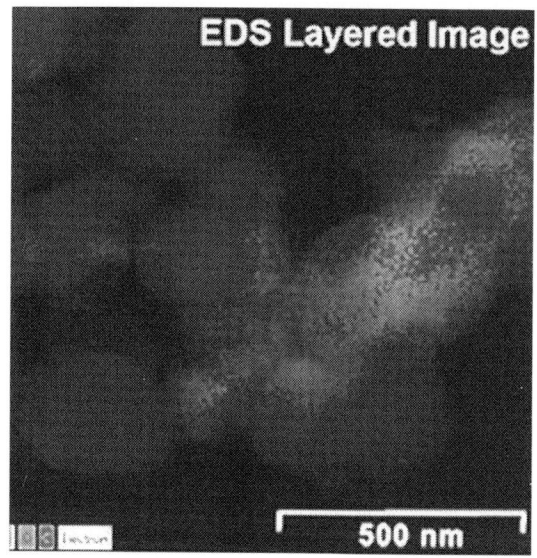

(b)

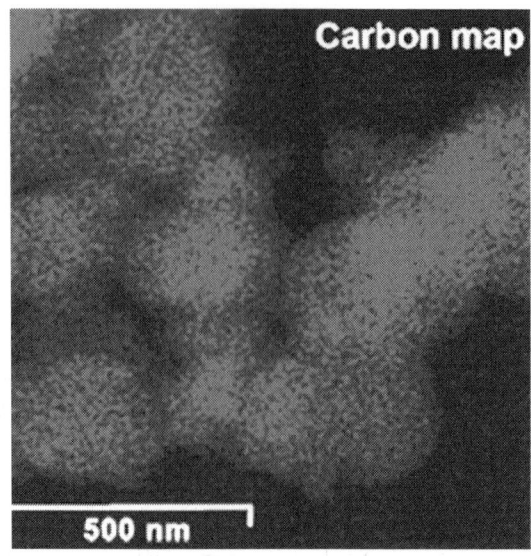

(c)

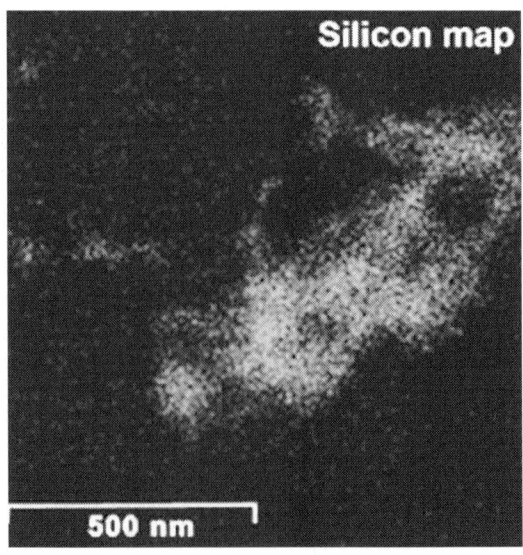

(d)

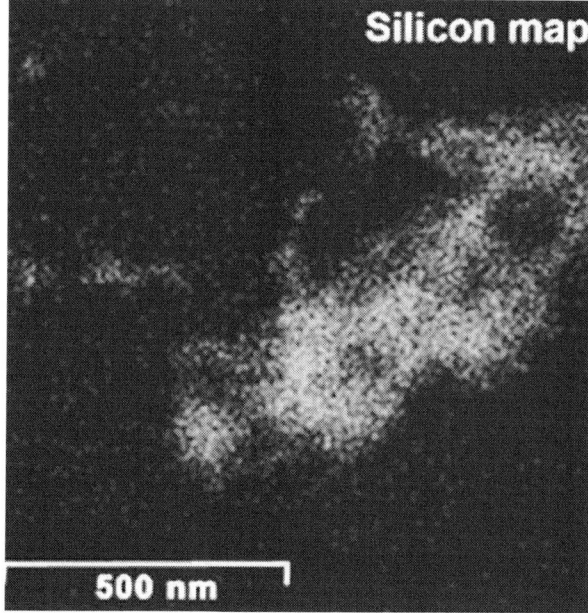

(e)

**Figure 4**: HAADF image and compositional maps of CSF-20.

Regarding the samples obtained by impregnation, it is quite clear that both phases also exist individually, i.e. discreet regions of amorphous $SiO_2$, and of soot-like carbon. The EDX maps show how the different particles were distributed. At a concentration of silica 10% (sample CSF-10), it is not to observe dominant placement of both types of particles. Varieties of cluster position are observed. Silica clusters are located both outside and inside carbon black clusters but it seems that more of them are outside, especially on the surface of carbon black clusters. In the samples with 20% concentration of silica (sample CSF-20), it is nearly not to see individual clusters of silica, particles of the latter have penetrated into the interior of the carbon black clusters. In more cases silica particles (as clusters) are inside the carbon black clusters, i.e. they are occluded, but some silica clusters are out of the carbon black clusters and are located as a coating of carbon black clusters. It seems that silica clustering is more often seen than silica coating of the carbon black clusters. According to the images, the most favorable sample is that with 10% silica. It creates the impression that the two phases are best blended. At the higher concentration (CSF-20), the better part of silica particles are located inside the carbon black clusters, which insulates them from contacts with elastomer macromolecules. That is confirmed also by the bright field TEM images of investigated samples made at different magnifications (Figures 5-8).

Indisputably the biggest difference established between hybrid fillers, obtained by both methods is the presence of zinc and sulfur in the fillers produced by pyrolysis in the form of zinc sulfide (Figure 1, Figure 2) whose presence is not found in the fillers obtained by other method. EDX spectra ascertain the availability of zinc sulphide (ZnS) in the fillers, obtained by pyrolysis (Figure 9). These spectra are taken from the three types of particles-ZnS, carbon and $SiO_2$, available in the hybrid fillers investigated. They all have some silica in; this is contamination (apart from the $SiO_2$ which actually has Si).

We can consider the presence of ZnS in the hybrid fillers investigated as a by-product, result of the reaction:

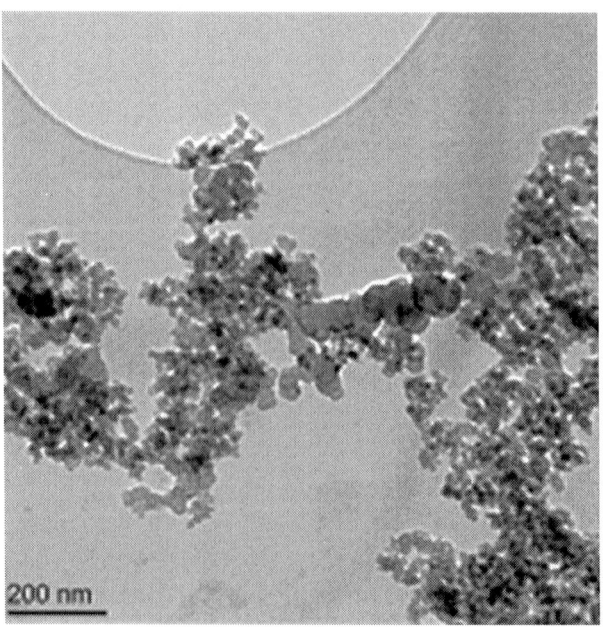

(a)

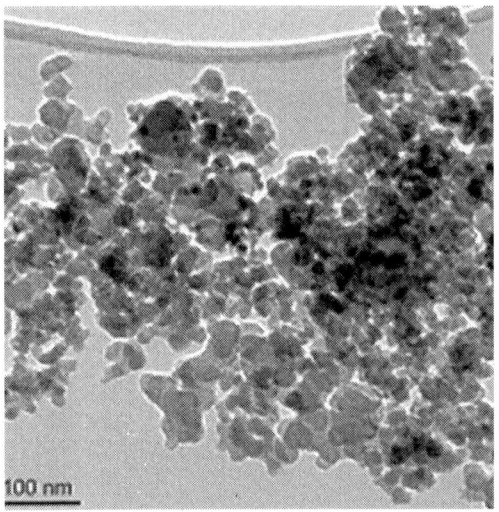

(b)

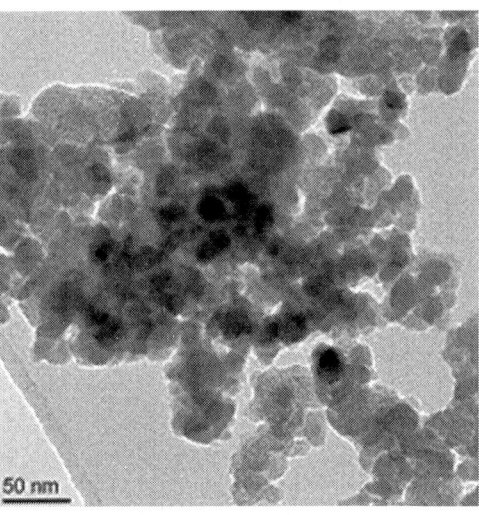

(c)

**Figure 5**: Bright field TEM images of CSF-1 at different magnifications.

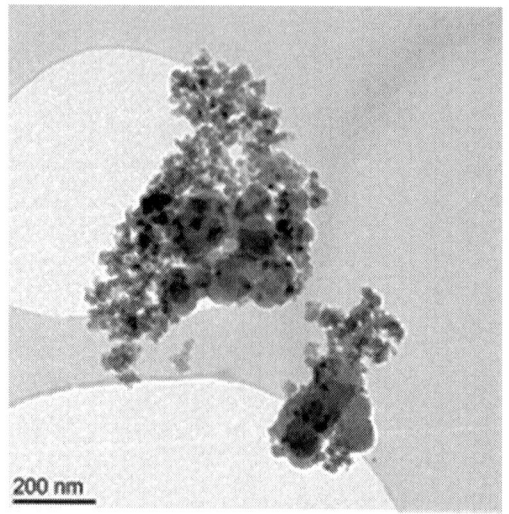

(a)

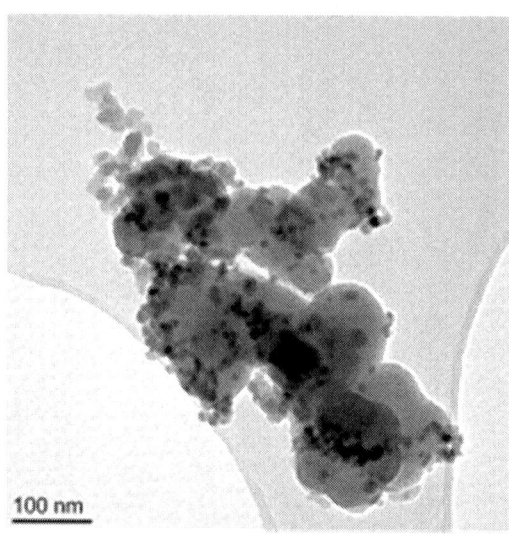

(b)

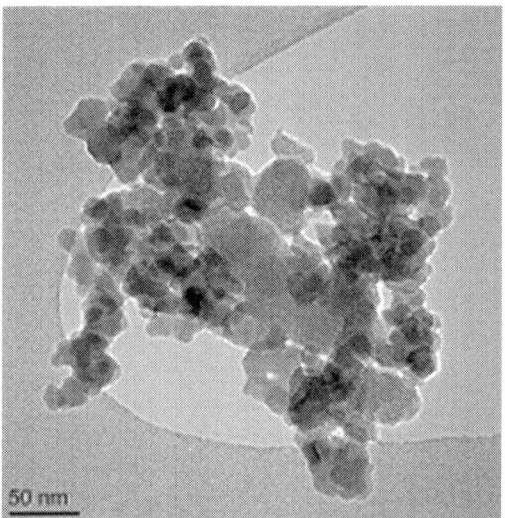

(c)

**Figure 6**: Bright field TEM images of CSF-2 at different magnifications.

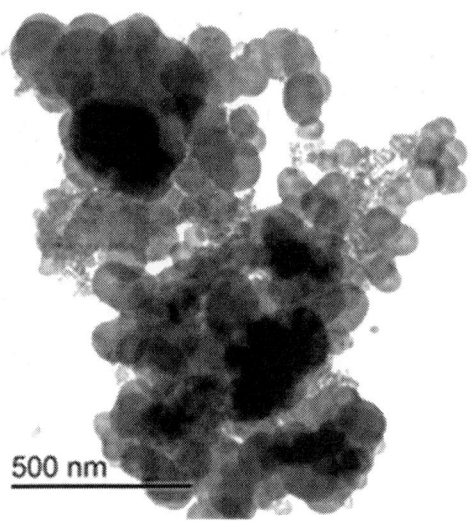

(a)

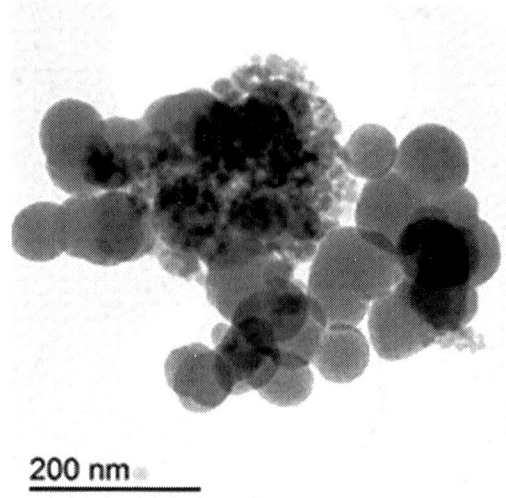

(b)

(c)

**Figure 7**: Bright field TEM images of CSF-10 at different magnifications.

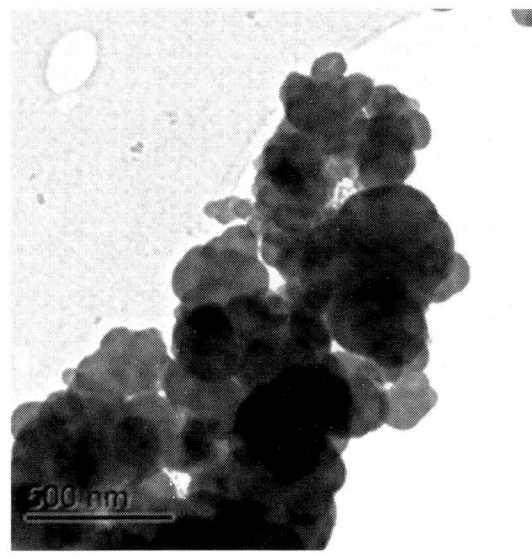

(a)

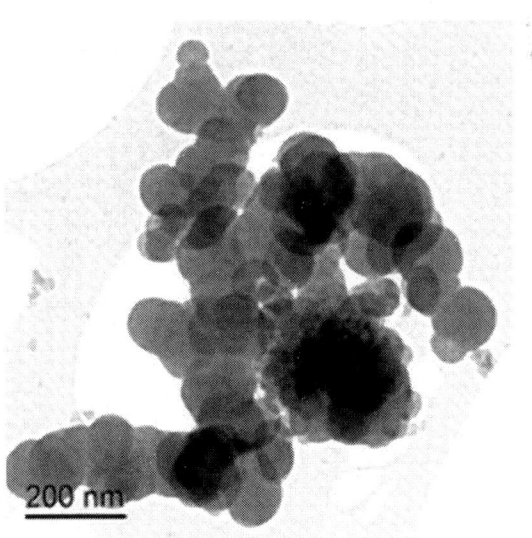

(b)

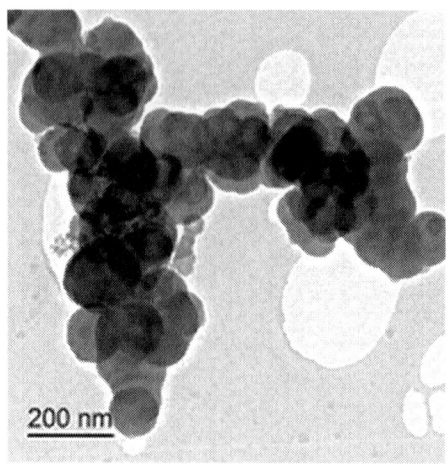

200 nm

(c)

**Figure 8**: Bright field TEM images of CSF-20 at different magnifications.

H$_2$S can be produced in the filler by reacting hydrogen gas (available in the air) with molten elemental sulfur over 450°C [21] during process of waste tires pyrolysis, carried out at 500°C and 700°C resp. for CSF-1 and CSF-2. Dohi and Horiuchi [22] investigating vulcanized rubber structures by energy-filtering transmission electron microscopy (EFTEM) and by high-angle annular dark field scanning transmission electron microscopy (HAADF-STEM) also observe numerous particles with the diameter of about 20 nm, distributed in the rubber matrix, which could not be observed by conventional TEM. Further examination based on electron energy-loss spectroscopy (EELS) and energy-dispersive X-ray (EDX) analysis indicated that those particles are composed of ZnS clusters with the sizes of approximately 3 - 5 nm. The authors consider that the particles are produced as a by-product in the vulcanization. They believe that the formation of ZnS clusters in a rubber network is one of the origins of rubber heterogeneities. There are data, available in the literature [23] that mechanical properties such as tensile and tear strength increases with increase in concentration of nano ZnS up to 7 phr of loading thereafter the value decreases, whereas hardness, and flame resistance increases with the dosage of fillers. Data confirming that ZnS has a heat stabilizing activity in synthetic rubber are also available [24] . On the other hand the presence of zinc oxide in the fillers (shown in Table 2) gives grounds to suppose that when using such fillers, vulcanization of mixtures could be held with a reduced amount of zinc oxide in their composition which is very important due to the ecological reasons. It is also obvious that the conditions of pyrolysis influence practically all the characteristics of hybrid filler, as can be seen from Table 1 andTable2 Research by STEM-EDX shows also their influence on the formation of silica aggregates, their size and placement of the carbon phase among them. It can be considered, that the conditions for obtaining of CSF-1 are more favorable from that point of view because the aggregates of the both phases with it, are not well isolated and their mutual penetration is at a higher level. Characteristics of CSF-1 and CSF-2, shown in Table 1 and Table 2 also give precedence to the first of them: its specific surface area is higher as well as iodine adsorption, ash content is lower, it contains bigger quantity of carbon phase etc.

Energy dispersive spectra on STEM of carbon black-silica filler obtained by impregnation are shown in Figures 10 and Figure 11.

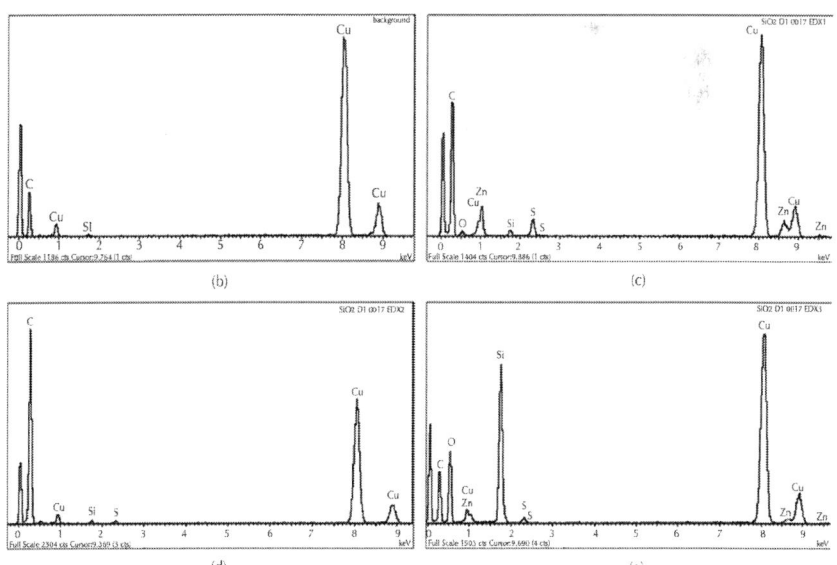

**Figure 9**: Energy dispersive spectra on STEM of carbon black-silica filler obtained by pyrolysis of waste green tires: (a) STEM image (annotated); (b) background; (c) EDX1 (ZnS); (d) EDX2 (carbon); (e) EDX3 ($SiO_2$).

EDX spectra indicate that there are zones that practically consist of carbon alone (3, 6), as well as areas in which carbon and silica coexist and are mixed on a very deep level. It is important to note that areas which contain only carbon are not noticed with the fillers obtained by pyrolysis, i.e. blending of both phases is really on a deeper level there. At that, there is tendency observed with increasing amount of silica in the sample, its quantity in these areas to enlarge too, judging for which by the increased intensity of silicon and oxygen picks (Spectra 7, 4).

It is also obvious that the quantity of introduced silica influences practically all the characteristics of hybrid filler, as it can be seen from Table 3 and Table4 Research by STEM-EDX shows also its influence on the formation of silica aggregates, their size and placement among the carbon black phase formations. It can be considered, that the silica concentration for obtaining of CSF-10 is more favorable from that point of view because the aggregates of the both phases are comparable and are not well isolated. Characteristics of CSF-10, shown in Table 1 and Table 2 also give precedence to this sample: its specific surface area is not so higher, while at the same time it is characterized by pores with the largest diameter, which creates conditions for the contact of the greatest number of elastomer macromolecules with filler and respectively the best strengthening effect.

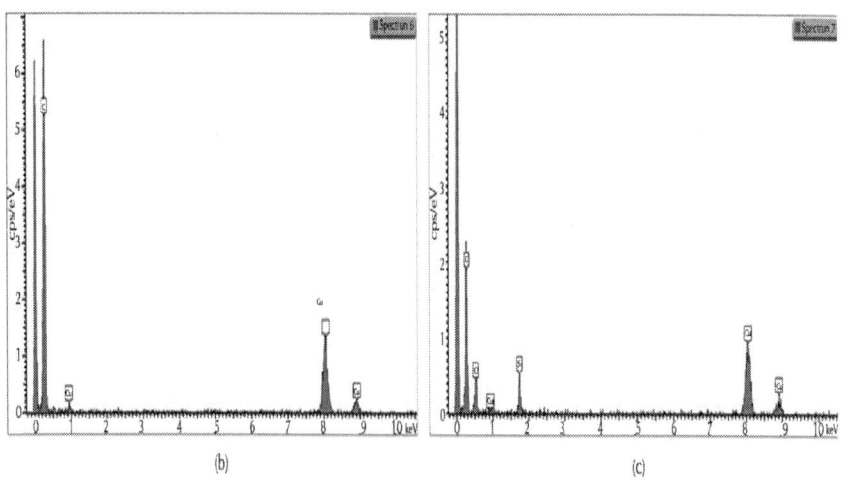

(b)    (c)

**Figure 10**: EDS on STEM of CSF-10 filler: (a) spectra positions; (b) spectrum 6; (c) spectrum 7.

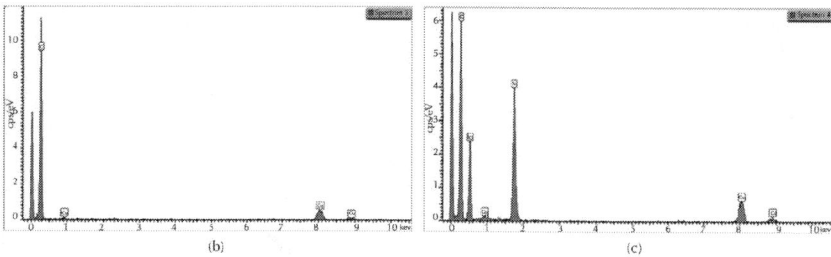

**Figure 11**: EDS on STEM of CSF-20 filler: (a) spectra positions; (b) spectrum 3; (c) spectrum 4.

# CONCLUSIONS

It has been compared the characteristics of hybrid fillers of the type of carbon, silica, obtained by two different methods: via pyrolysis and by impregnation. It was found that the characteristics of the used fillers differ substantially; in some cases they are even opposite, due to contrary approaches for their obtaining.

It is obvious that the creation of dual phase fillers according to the proposed methods is one way, allowing for changes in the desired direction of all important factors affecting the improvement of elastomers: size and specific surface area of the particles, morphology, structure, degree of irregularity, and surface activity.

In most aspects carbon-silica fillers differ from the traditional fillers used as reinforcing agents in the rubber industry. The possibilities they discover to control the interactions "elastomer-filler" and "filler-filler" and from here a number of operational properties of final products (dynamic, mechanical, etc.) can be considered as very perspective.

# ACKNOWLEDGEMENTS

The present research is a result of an international collaboration program between University of Tabuk, Tabuk, Kingdom of Saudi Arabia and the University of Chemical Technology and Metallurgy, Sofia, Bulgaria. The authors gratefully acknowledge the financial support from the University of Tabuk.

Authors wish also to thank to Dr. Antonia Dimitrova, SPEME, University of Leeds, Leeds LS2 9JT, UK for her assistance in STEM-EDX analyses, as well as to Prof. Ljutzkan Ljutzkanov, Institute of Chemical Engineering, Bulgarian Academy of Sciences, Sofia, Bulgaria, for the preparation of the carbon-silica fillers by pyrolisis.

# REFERENCES

1.  Frohlich, J., Niedermeier, W. and Luginsland, H.-D. (2005) The Effect of Filler-Filler and Filler-Elastomer Interaction on Rubber Reinforcement. Composites: Part A, 36, 449-460.

2.  Donnet, J.-B. and Custodero, E. (2013) Reinforcement of Elastomers by Particulate Fillers. In: Mark, J.E., Erman, B. and Roland, C.M., Eds., The Science and Technology of Rubber, 4th Edition, Elsevier, Amsterdam, 383-416.http://dx.doi.org/10.1016/B978-0-12-394584-6.00008-X

3.  Mahmud, K., Wang, M.-J. and Francis, R.A. (1998) Elastomeric Compound Incorporating Silicon-Treated Carbon Black. US Patent No. 5830930.

4.  Wang, M.J., Kutsovsky, Y., Zhang, P., Mehos, G., Murphy, L.J. and Mahmud, K. (2002) Using Carbon-Silica Dual Phase Filler Improve Global Compromise between Rolling Resistance, Wear Resistance and Wet Skid Resistance for Tires. Kautschuk Gummi Kunststoffe, 55, 33-40.

5.  Wang, M.J., Mahmud, K., Murphy, L.J. and Patterson, W.J. (1998) Carbon-Silica Dual Phase Filler—A New Generation Reinforcing Agent for Rubber. Kautschuk Gummi Kunststoffe, 51, 348-360.

6.  Kolev, D., Ljutzkanova, R. and Abadjiev, S. (2005) Method and Installation for Pyrolysis of Tires. Bulgarian Patent 65901 B1.

7.  Kolev, D., Ljutzkanova, R. and Abadjiev, S. (2008) Method and Installation for Pyrolysis of Tires. European Patent EP 1879978 A1.

8.  Isayev, A. (2013) Recycling of Rubbers. In: Erman, B., Mark, J. and Roland, C., Eds., The Science and Technology of Rubber, 4th Edition, Elsevier, Amsterdam, 753-755.http://dx.doi.org/10.1016/B978-0-12-394584-6.00020-0

9. Ivanov, M., Mihaylov, M. and Ljutzkanov, L. (2010) Silica Obtained via Pyrolysis of Waste "Green" Tyres—A Perspective Filler for Rubber Industry. Kautschuk Gummi Kunststoffe, 63, 303-307.

10. Ivanov, M. and Mihaylov, M. (2011) Silica Obtained via Pyrolysis of Waste "Green" Tyres—A Filler for Tyre Tread Rubber Blends. Journal of Elastomers and Plastics, 43, 303-316.http://dx.doi.org/10.1177/0095244311398636

11. Al-Hartomy, O.A., Al-Ghamdi, A.A., Al Said, S.A., Dishovsky, N., Mihaylov, M., Ivanov, M. and Ljutzkanov, L. (2013) Effect of the Carbon-Silica Reinforcing Filler Obtained from the Pyrolysis-cum-Water Vapour of Waste Green Tyres upon the Properties of Natural Rubber Based Composites. Progress in Rubber, Plastics and Recycling Technology, in press.

12. Al-Hartomy, O.A., Al-Ghamdi, A.A., Al-Said, S.A., Dishovsky, N., Mihaylov, M., Ivanov, M. and Ljutzkanov, L. (2014) Effect of the Solid Product Obtained by Pyrolysis of Waste Green Tires on the Properties of Epoxidized Natural Rubber Based Composites. International Review of Chemical Engineering, in press.

13. Al-Hartomy, O.A., Al-Ghamdi, A.A., Al-Said, S.A., Dishovsky, N., Mihaylov, M., Ivanov, M. and Ljutzkanov, L. (2013) Influence of the Carbon-Silica Reinforcing Filler Obtained via Pyrolysis of Waste "Green" Tyres on the Properties of EPDM Based Composites. Kautschuk Gummi Kunststoffe, in press.

14. Marsh, H., Heintz, E. and Rodrigues-Reinoso, F. (1997) Introduction to Carbon Technologies. University of Alicante, Alicante.

15. Malinova, P., Nikolov, R., Dishovsky, N. and Lakov, L. (2004) Modification of Carbon-Containing Fillers for Elastomers. Kautschuk Gummi Kunststoffe, 57, 443-445.

16. Al-Hartomy, O.A., Al-Ghamdi, A.A., Al-Said, S.A., Dishovsky, N., Malinova, P. and Nikolov, R. (2013) Obtaining of Carbon-Silica Dual Phase Filler by Impregnation Method and Investigation of Its Influence on the Properties of SBR Based Composites. Journal of Materials Design and Application, in press.

17. ISO 15651/1-91.

18. Bulgarian State Standard 9665:1176.

19.  Brunauer, S., Emmett, P.H. and Teller, E. (1938) Adsorption of Gases in Multimolecular Layers. Journal of the American Chemical Society, 60, 309-319.http://dx.doi.org/10.1021/ja01269a023

20.  ISO 1125:1999.

21.  Ullmann's Encyclopedia of Chemical Industry, 1999-2014. John Wiley & Sons Inc.

22.  Dohi, H. and Horiuchi, S. (2007) Heterogeneity of a Vulcanized Rubber by the Formation of a ZnS Clusters. Polymer, 48, 2526-2530.http://dx.doi.org/10.1016/j.polymer.2007.03.004

23.  Ramesan, M.T., Nihmath, A. and Francis, J. (2013) Preparation and Characterization of Zinc Sulphide Nanocomposites Based on Acrylonitrile Butadiene Rubber. Proceeding of International Conference on Recent Trends in Applied Physics & Material Science RAM 2013, Bikaner, 1-2 February 2013, 255-256.

24.  David, S., Fritzen, P., Heiming, L. and Rentschler, T. (2010) Plastic Comprising ZnS. US Patent No. 8383712.

# Chapter 6

# Thermal Conductivity Enhancement of Epoxy by Hybrid Particulate Fillers of Graphite and Silicon Carbide

Krishnamachar Srinivas[1] and Mysore Siddalingappa Bhagyashekar[2]

[1]Department of Mechanical Engineering, Don Bosco Institute of Technology, Bangalore, India

[2]Department of Mechanical Engineering, Raja Rajeshwari College of Engineering, Bangalore, India

## ABSTRACT

This paper reports thermal conductivity studies carried out on room temperature cure (RT) epoxy resin (LY556 + HY951) containing three different particulate fillers such as Graphite (Gr) a soft material, Silicon carbide (SiC) a hard material and a hybrid graphite & silicon carbide (Gr-SiC). The weight fractions of three different fillers were varied from

10 wt% to 40 wt% in steps of 10%. Increased filler fraction increases the thermal conductivity of epoxy composites for all the three types of fillers. The results show that the synergic effect of hybrid filler (Gr-SiC) improves the thermal conductivity of epoxy composites compared to that of Graphite or Silicon carbide. The improvement in thermal conductivity for the epoxy hybrid composite containing 20% SiC, 20% Gr and 60% epoxy is 136% when compared with neat epoxy. Significant improvement in the thermal conductivity is observed for 40% filled epoxies. The experimental results were compared with analytical models such as Maxwell, Hashin, Hamilton-Crosser, Nielsen, and Cheng-Vachon. The predicted thermal conductivity by analytical models is in agreement with experimental results.

# INTRODUCTION

Polymers in general have low thermal conductivity due to low atomic density, chemical bonding, complex crystal structure [1] and anharmonicity in molecular vibrations [2]. The heat transfer in nonmetals occurs by phonon or lattice vibrations causing thermal resistance [3]. Epoxy as a matrix has a wide applications in the field of engineering. They have some of the favorable properties such as compatibility with many materials, higher thermal and chemical resistance cheaper than other thermosets. However, due to its cross linked structure, epoxies have disadvantages like brittleness, low impact strength, low wear resistance [4] . Epoxies are poor conductors of heat and current. Hence to improve their thermal conductivity, one of the routes is to reinforce particulate fillers. Excellent conductive properties of epoxies have applications in miniature devices managing heat transfer. To improve thermal conductivity, the thermal resistance has to be minimized. A good interfacial adhesion by modifying the filler surface reduces thermal resistance [5]. Further, when the thermal conductivity of the filler is more than 100 times the thermal conductivity of the polymer matrix, there will not be substantial increase in thermal conductivity of the composite [3] . To overcome this, the aspect ratio of the filler has to be increased so as to create conductive paths [6] [7] . The conductive paths can be created by increasing the volume fraction of fillers, increasing the size of particles and use of hybrid fillers. The fillers with large aspect ratio are capable of creating a conductive network [8] [9]

. Hence, the present study looks into the modification of epoxy, by incorporation of particulates of Gr, SiC and hybrid (Gr-SiC) at various weight fractions to improve thermal conductivity. The Gr is considered as one of the best conductive filler due to its ability to disperse in a matrix, low cost and good thermal conductivity [1] . SiC microfiller is a choice material for high temperature and high power applications, as it has high thermal conductivity and low thermal expansion coefficient [10] .

# EXPERIMENTAL DETAILS

## Materials Selecting a Template

Micro particulates of Graphite (Gr) supplied by SD fine chem ltd, India having size less than 60 μm and particulates of Silicon carbide (SiC) supplied by Gran Silica India, having size less than 20 μm (Figure 1) were used as fillers. The host matrix is an epoxy resin of type LY556 (RT cure) and an amine hardener HY951 supplied by Huntsman India Ltd. The resin and hardener were mixed in the weight ratio 10:1.

## Specimen Preparation and Characterization

A metallic mould was fabricated to give specimens of 2 mm thick and 25 mm diameter which is coin shaped as shown in Figure 2. Before casting, the mould was cleaned thoroughly and a gel coat was applied to the mould. The particulates were preheated to remove any moisture present in it. The predetermined amounts of fillers were dispersed in epoxy resin, stirring continuously till the time of pouring. The cast specimens were allowed to cure in the mould for 24 hours at room temperature. The room temperature cured specimens were post cured for the following schedule of 50°C for 30 minutes, 70°C for 60 minutes and 85°C for 120 minutes. To begin with, unfilled epoxy was cast first. Further, Gr-epoxy, SiC-epoxy and hybrid (Gr-SiC) epoxy composites were cast. The weight fractions of particulates of Gr and SiC were varied from 10% to 40% (maximum) in steps of 10%. In the next step, hybrid (Gr-SiC) filled epoxy composites were cast containing the filler weight fraction as given below.

(a)

(b)

**Figure 1**: SEM Photographs of particulates of Gr and SiC. (a) SiC particles; (b) Gr particles

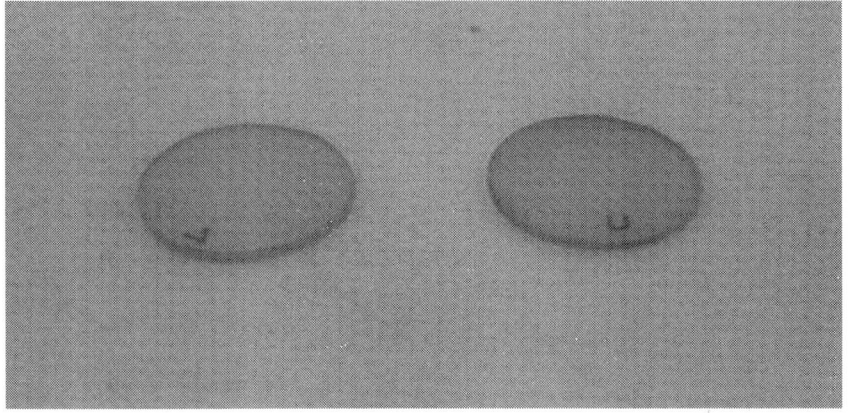

(a)

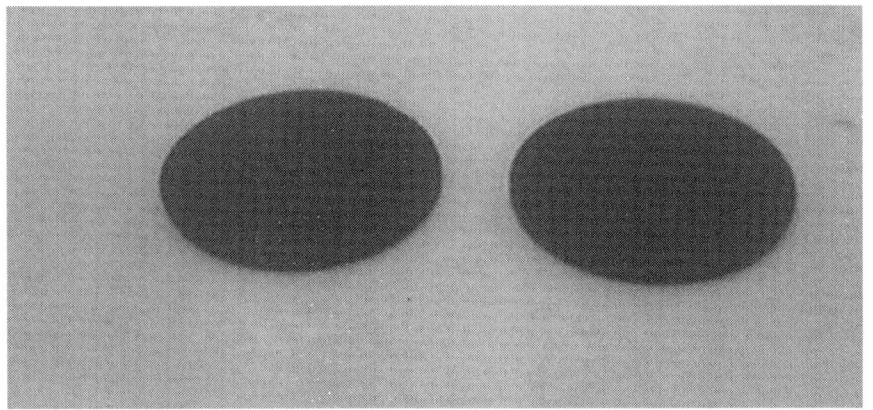

(b)

**Figure 2**: Cast specimens of epoxy and (Gr-SiC) epoxy hybrid composite. (a) Unfilled epoxy; (b) (Gr-SiC) Hybrid epoxy composite.

Gr fraction kept at 5%, SiC fraction varied from 5% to 35% in steps of 10%.

Gr fraction kept at 10%, SiC fraction varied from 10% to 30% in steps of 10%.

SiC fraction kept at 5%, Gr fraction varied from 5% to 35% in steps of 10%.

SiC fraction kept at 10%, Gr fraction varied from 10% to 30% in steps of 10%.

Both SiC and Gr fraction varied equally up to 40% in steps of 10%.

The different types of cast composites along with their codes are presented in Table 1. For ex composite 5S35G-Ep represents particulate content of 5% SiC, 35% Gr and 60% epoxy. Similarly a composite having code 10G30S-Ep has 10% Gr, 30% SiC and 60% epoxy.

# Measurement of Thermal Conductivity

Thermal conductivity measurements were carried out by an instrument developed by Indian national academy of science. The schematic diagram of the set up is shown in Figure 3. The thermal conductivity measurements were based on steady state method. One directional heat flow was considered for measurement of thermal conductivity.

# Description of the Apparatus

A heater made of six 120-Ohm 1/4 W resistors in parallel is placed in the copper cup of 25 mm diameter. It is connected to a constant current source. Two cast specimens are clamped between the copper cup and two square copper plates. Two junctions of a copper constantan thermocouple are fixed to upper half of the copper heater cup and to the upper square plate. Another pair of junctions of the thermocouple is fixed to bottom edge of the bottom half of the heater cup and bottom square plate. The leads from the thermocouple are connected to the differential amplifier. The specimens are fixed in between the square plate and the copper cup. The setup is connected to a constant current source. The outputs from the thermocouples are connected to a DC differential amplifier. Initially the current is set at 140 mA. The steady state is obtained after 45 mins. The output voltage is measured on a digital multimeter for thermocouple 1 and the process is repeated for thermocouple 2. Further the current is increased to 160 mA and the measurements are repeated. The measurements are repeated for five different current settings and two specimens were used for each experiment.

The experimental thermal conductivity is evaluated using the equation

$$Q = -KA\left[\frac{\Delta T}{d}\right]$$

(1)

Where, K is the thermal conductivity of composite in W/mK;

Q is the heat supplied in watts;

A is the area of the specimen in m²;

d is the thickness of the specimen in m.

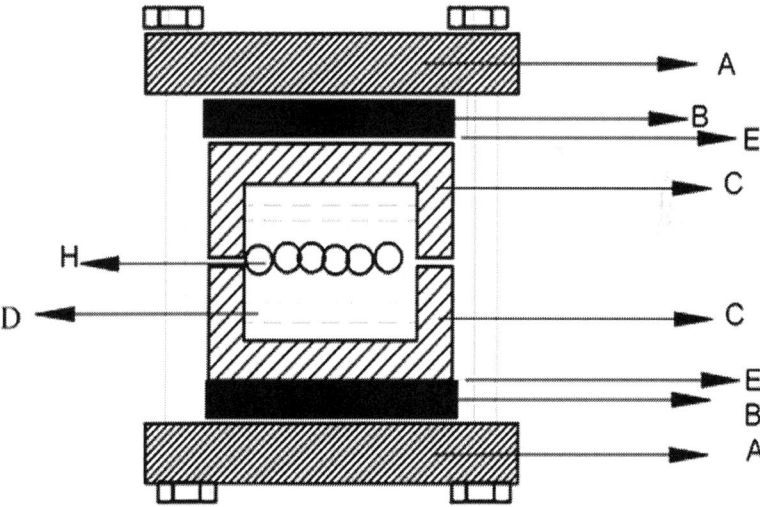

**Figure 3**: Schematic Diagram of the set up. A: Copper base plate each 50 mm square 5 mm thick; B: Specimen 25 mm dia and 2 mm thick; C: Two halves of a copper cup 25 mm dia, outer height 5 mm and inner height 3 mm; D: Aluminium foils pads; H: Heater made up of six 120 ohm 1/4 W resistors in parallel; E: Cu-Constantan thermocouples.

**Table 1:** Thermal conductivity of epoxy and its composites

| | Code | %Ep | %SiC | %Gr | Expt (W/mK) | Series (W/mK) | Maxwell (W/mK) | Hashin (W/mK) | Neilson (W/mK) |
|---|---|---|---|---|---|---|---|---|---|
| Epoxy and 10% Filled epoxy | Ep | 100 | 0 | 0 | 0.30 | 0.30 | 0.30 | 0.30 | 0.30 |
| | 10G-Ep | 90 | 0 | 10 | 0.32 | 0.33 | 0.39 | 0.37 | 0.39 |
| | 10S-Ep | 90 | 10 | 0 | 0.33 | 0.33 | 0.39 | 0.37 | 0.38 |
| | 5S5G-Ep | 90 | 5 | 5 | 0.33 | 0.39 | 0.37 | 0.38 | 0.33 |
| 20% Filled epoxy | 20G-Ep | 80 | 0 | 20 | 0.46 | 0.37 | 0.48 | 0.47 | 0.51 |
| | 20S-Ep | 80 | 20 | 0 | 0.41 | 0.37 | 0.49 | 0.45 | 0.49 |
| | 5S15G-Ep | 80 | 5 | 15 | 0.46 | 0.48 | 0.45 | 0.52 | 0.40 |
| | 5G15S-Ep | 80 | 15 | 5 | 0.46 | 0.48 | 0.45 | 0.52 | 0.40 |
| | 10S10G-Ep | 80 | 10 | 10 | 0.51 | 0.48 | 0.45 | 0.50 | 0.37 |
| 30% Filled epoxy | 30G-Ep | 70 | 0 | 30 | 0.54 | 0.43 | 0.57 | 0.59 | 0.7 |
| | 30S-Ep | 70 | 30 | 0 | 0.51 | 0.43 | 0.58 | 0.56 | 0.63 |
| | 5S25G-Ep | 70 | 5 | 25 | 0.52 | 0.57 | 0.56 | 0.67 | 0.43 |
| | 5G25S-Ep | 70 | 20 | 5 | 0.52 | 0.57 | 0.56 | 0.67 | 0.43 |
| | 10S20G-Ep | 70 | 10 | 20 | 0.56 | 0.54 | 0.55 | 0.72 | 0.45 |
| | 10G20S-Ep | 70 | 20 | 10 | 0.54 | 0.54 | 0.55 | 0.64 | 0.45 |
| | 15G15S-Ep | 70 | 15 | 15 | 0.58 | 0.60 | 0.59 | 0.69 | 0.45 |

| 40% Filled epoxy | | | | | | | | |
| --- | --- | --- | --- | --- | --- | --- | --- | --- |
| 40G-Ep | 60 | 0 | 40 | 0.64 | 0.50 | 0.66 | 0.73 | 1.04 |
| 40S-Ep | 60 | 40 | 0 | 0.68 | 0.50 | 0.70 | 0.70 | 0.82 |
| 5S35G-Ep | 60 | 5 | 35 | 0.64 | 0.66 | 0.70 | 0.86 | 0.50 |
| 5G35S-Ep | 60 | 35 | 5 | 0.62 | 0.66 | 0.70 | 0.86 | 0.50 |
| 10S30G-Ep | 60 | 10 | 30 | 0.62 | 0.66 | 0.7 | 1.02 | 0.53 |
| 10G30S-Ep | 60 | 30 | 10 | 0.61 | 0.66 | 0.70 | 0.84 | 0.53 |
| 20G20S-Ep | 60 | 20 | 20 | 0.71 | 0.71 | 0.70 | 0.90 | 0.50 |

## Theoretical Models

The following analytical models were used to predict the thermal conductivity of epoxy composites [11] [12]. The models considered were Maxwell equation, cylinder assemblage model developed by Hashin Nielsen's equation and series equation, which is an inverse rule of mixtures which gives lower bound values of thermal conductivity for particulate filled composites [12].

a)  Maxwell Equation

$$\frac{K}{Kc} = 1 + \left[\frac{3(Kd - Kc)}{(Kd + 2Kc)}\right]\Phi$$

$$(2)$$

Where K, Kc and Kd are the thermal conductivities of composites, matrix and fillers $\Phi$ is the volume fraction of filler.

b)  Cylinder assemblage model: Cylinder Assemblage model was developed by Hashin

$$K_T = K_{TM}\left\{K_{TM}V_M + K_{TF}\left(1 + V_F\right)/K_{TM}\left(1 + V_F\right) + K_{TF}V_M\right\}$$

$$(3)$$

Where, $K_T$, $K_{TM}$ and $K_{TF}$ are thermal conductivities of composite, matrix and filler.

$V_F$, $V_M$ are the volume fraction of filler and matrix.

c)  Nielsen's Equation

$$\frac{K_c}{K_m} = \frac{1 + AB\varphi}{1 - \phi B\varphi}$$

Where

$$B = \frac{k_f / k_m - 1}{k_f / k_m + A} \quad \text{and} \quad \phi = 1 + \frac{(1 - \varphi_{max})\varphi}{\varphi^2_{max}}$$

(4)

where A is the geometry of the particles = 1.5 and $\phi_{max}$ (maximum packing fraction)

d) Inverse rule of mixtures/Series

$$\frac{1}{K_T} = \frac{V_f}{K_f} + \frac{V_m}{K_m}$$

(5)

The above equation is derived from rule of mixtures and does not consider voids.

# RESULTS AND DISCUSSION

## Effect of Gr/SiC on Thermal Conductivity of Gr-Epoxy/SiC-Epoxy Composites

The experimental thermal conductivity and predicted by analytical models of all types of epoxy composites is listed in Table 1. Figure 4(a) and Figure 4(b) presents the thermal conductivity of epoxy composites filled with particulates of Gr and SiC. Neat epoxy exhibits a thermal conductivity of 0.30 W/mK. For filled Gr/SiC epoxy, up to 10% filler fraction, the rise in thermal conductivity is not significant. The epoxy composites containing 10% fillers, 10G-Ep and 10S-Ep exhibits a thermal conductivity of 0.32 W/mK and 0.33 W/mK respectively which is an insignificant improvement over neat epoxy. This may be due to low filler fraction and hence effective conductive paths are not created in the composite. Further addition of particulates of graphite increases thermal conductivity of graphite epoxy composites linearly.

The thermal conductivity increases to 0.46 W/mK and 0.54 W/mK for 20G-Ep and 30G-Ep respectively. Maxima of 0.64 W/mK was observed for 40G-Ep which is an improvement of 113% over unfilled epoxy. The effect of adding SiC to epoxy is shown in Figure 4(b). The SiC-epoxy composites have trends similar to Gr-epoxy composites. However, 20S-Ep and 30S-Ep have thermal conductivity of 0.41 W/mK and 0.51 W/mK respectively which is marginally lower than Gr-epoxy under similar loading. However 40S-Ep, exhibits a thermal conductivity of 0.68 W/mK. This suggests that even though the thermal conductivity of SiC (280 W/mK) is higher than Gr (85 W/mK), SiC-Ep composites have lower thermal conductivity than Gr-Ep composites. The factors such as larger particle size of Gr (20 - 60 µm) (Figure 1(b)) irregularity in shape, easy dispersibility [11] favors enhancement of thermal conductivity, where as SiC particulates are smaller in size (6 - 12 µm) uniform (Figure 1), are not beneficial in creating conductive paths [8] [9] .

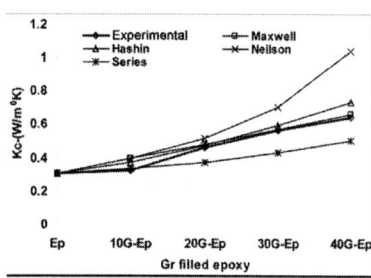 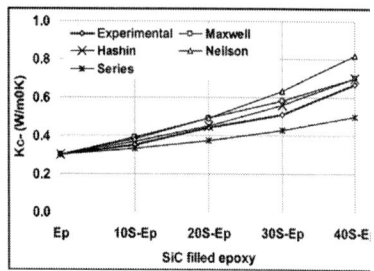

**Figure 4**: Thermal conductivity of Gr/SiC filled epoxy. (a) Gr filled epoxy; (b) SiC filled epoxy.

Comparing thermal conductivity experimental with thermal conductivity predicted by analytical equations, it is seen that, series equation predicts lower thermal conductivity for all filler loadings where as Hashin and Maxwell equations are closer to experimental values. The prediction by Nielsen's equation is in agreement with experimental value up to 20% filler loading and deviates after 30% filler fraction. This is in agreement with [3] [5] , where the Nielsen's estimation at lower volume fraction matches the experimental values and tends to over estimate the thermal conductivity of composites at higher volume fraction.

# Effect of Gr on Thermal Conductivity of Hybrid (Gr-SiC)-Epoxy Composites

The effect of varying Gr on hybrid (Gr-SiC) epoxy composites is shown in Figure 5(a) and Figure 5(b). Figure 5(a) shows the thermal conductivity of epoxy composites, where SiC was unvaried at 5% for all composites, Gr varied from 5% to 35% in steps of 10%. Increased Gr fraction increases the thermal conductivity of hybrid (Gr-SiC) epoxy composites. Thermal conductivity increases to 0.46 W/mK for 5S15G-Ep, 0.52 W/mK for 5S25G-Ep and 0.64 W/mK for 5S35G-Ep composite.

The effect of Gr where SiC is held at 10% and Gr is varied from 10% to 30% is shown in Figure 5(b). These composites present a trend similar to previous compositions. However, these composites exhibits marginally lower conductivity values than composites having 5% SiC unvaried composites. Maximum thermal conductivity of 0.62 W/mK was observed for 10S30G-Ep. The SEM of the surfaces of 5S35G-Ep and 10S30G-Ep is shown in Figure 6(a) and Figure 6(b) respectively. The few bright spots seen in the images are particulates of SiC.

# Effect of SiC on Thermal Conductivity of Hybrid (Gr-SiC)-Epoxy Composites

Figure 7(a) and Figure 7(b) illustrates the effect of varying SiC and holding Gr constant at 5%, 10% respectively on hybrid (Gr-SiC) epoxy composites. Thermal conductivity increases linearly for all type of composites. The thermal conductivity increases to 0.52 W/mK for 5G25S-Ep and 0.62 W/mK for 5G35S-Ep. Similarly 10G20S-Ep and 10G30S-Ep exhibits thermal conductivity of 0.54 W/mK and 0.62 W/mK respectively. The SEM images of 5G35S-Ep and 10G30S-Ep composites are shown in Figure 8(a) and Figure 8(b). As SiC fraction is more for these composites, more bright spots are seen, indicating well distribution of SiC.

# Effect of Equal Fraction of Gr and SiC on Thermal Conductivity of Hybrid (Gr-SiC)-Epoxy Composites

The effect of hybrid (Gr-SiC) fillers on epoxy where both fillers are equally varied is shown in Figure 9. The thermal conductivity of 10G10S-Ep is found to be 0.51 W/mK which is highest among 20% filled composites. Similarly 15G15S-Ep exhibits a thermal conductivity of 0.58 W/mK, highest among 30% filled epoxy. Similarly the thermal conductivity of 20G20S-Ep is 0.71 W/mK, highest among 40% filled composites. Further, it is also observed that for the same filler loadings, the thermal conductivity varies. As filler loading increases, number of interfaces increases and thermal conductivity in a polymer composite is dependent on good interfacial adhesion. For a poor interfacial adhesion phonon scattering increases, and decreases the thermal conductivity. Further different particle sizes vary the interfaces along the heat transfer path [13] . From the above it is evident that synergy exists between hybrid (Gr-SiC) when both the fillers are loaded equally.

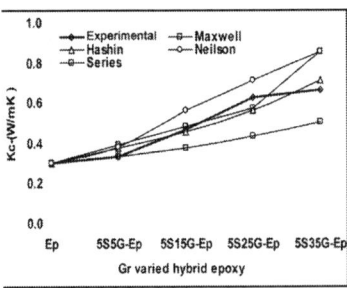 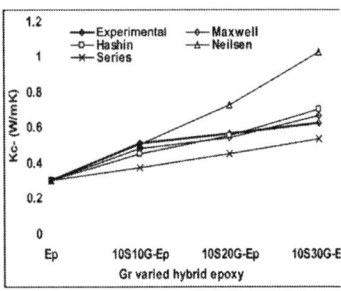

**Figure 5**: Thermal conductivity of graphite rich hybrid (Gr-SiC) epoxy composites. (a) SiC constant at 5%, Gr varied from 5% - 35%; (b) SiC constant at 10%, Gr varied from 10% - 30%.

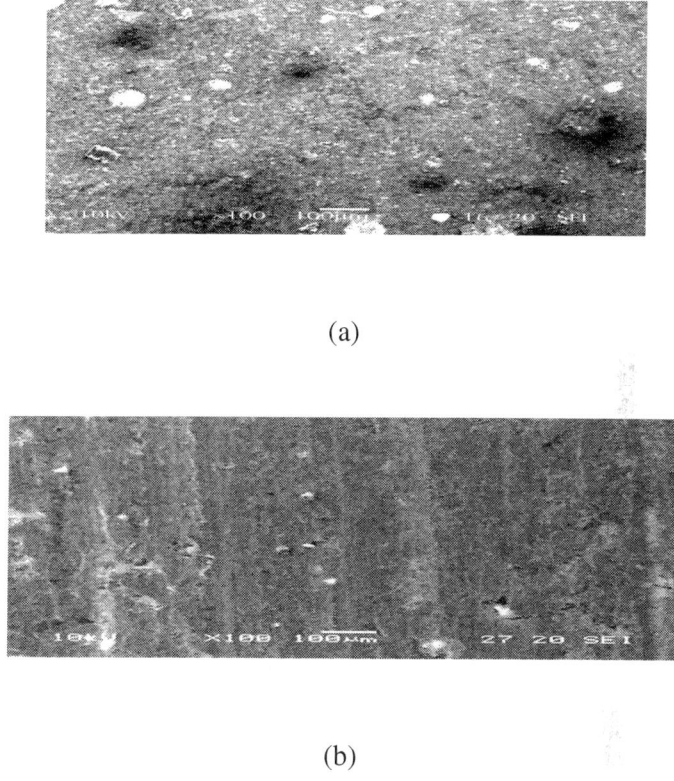

(a)

(b)

**Figure 6**: SEM images of surface of graphite rich hybrid (Gr-SiC) epoxy composites. (a) 5S35G-Ep; (b) 10S30G-Ep

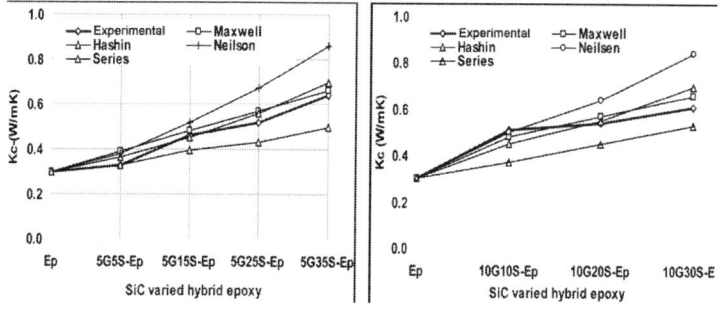

**Figure 7**: Thermal conductivity of SiC rich hybrid (Gr-SiC) epoxy composites. (a) Gr constant at 5%, SiC varied from 5% - 35%; (b) Gr constant at 10%, SiC varied from 10% - 30%.

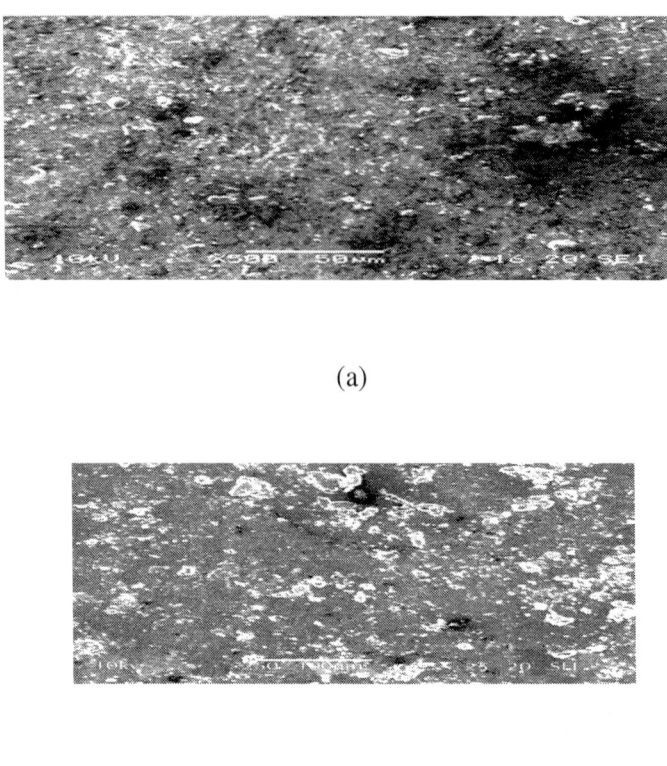

(a)

(b)

**Figure 8**: SEM images of surface of graphite rich hybrid (Gr-SiC) epoxy composites. (a) 5G35S-Ep; (b) 10G30S-Ep

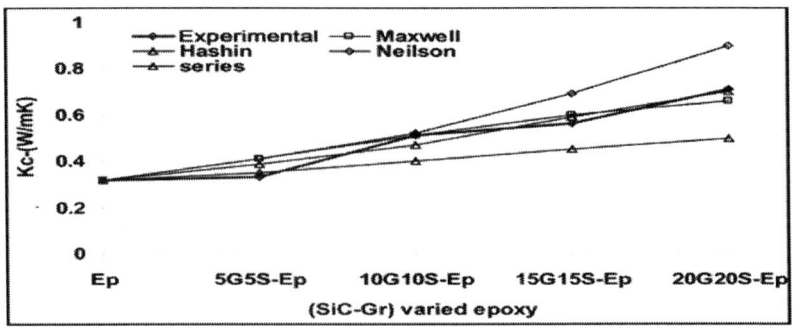

**Figure 9**: Effect of fillers on thermal conductivity of epoxy composites. Both Gr and SiC varied equally.

# CONCLUSIONS

Thermal conductivity increases linearly with increased filler fraction for all composite systems irrespective of fillers. Thermal conductivity show significant increase after 20% filler fraction. Comparing Gr-Ep and SiC-Ep composites, for the same filler fraction, Gr-Ep composites have higher thermal conductivity than SiC-Ep composites except for 40S-Ep which has higher thermal conductivity than 40G-Ep. Hybrid composites having equal proportion of filler fraction i.e. 10G10S-Ep, 15G15S-Ep and 20G20S-Ep have thermal conductivity highest in 20%, 30% and 40% filled epoxy composites. The hybrid composite 20G20S-Ep has a thermal conductivity of 0.71 W/mK which is highest among all the composite considered in this study and is an improvement of 136% over a neat epoxy. The experimental results were compared with the analytical models. The experimental results exhibit good correlation with Maxwell's equation.

# ACKNOWLEDGEMENTS

The authors thank the management of Don Bosco Institute of Technology, RajaRajeswari College of Engineering Bangalore and Indian Academy Of Science. Bangalore.

# REFERENCES

1.    Ebadi-Dehaghani, H. and Branch, M.N.S. (2012) Thermal Conductivity of Nanoparticles Filled Polymers Islamic Azad University Iran Smart Nanoparticles Technology. 519-540.

2.    Tsekmes, I.A., Kochetov, R., Morshuis, P.H.F. and Smit, J.J. (2013) Thermal Conductivity of Polymeric Composites: A Review. 2013 IEEE International Conference on Solid Dielectrics, Bologna, 30 June-4 July 2013, 678-681.

3.    Leea, G.-W., Parka, M., Kima, J., Leeb, J.I. and Yoonb, H.G. (2006) Enhanced Thermal Conductivity of Polymer Composites Filled with Hybrid Filler. Composites: Part A, 37, 727-734. http://dx.doi.org/10.1016/j.compositesa.2005.07.006

4.    Wang, M.C., Zhang, Z.G. and Sun, Z.J. (2009) The Hybrid Model and Mechanical Properties of Hybrid Composites Reinforced with Different Diameter Fibers. Journal of Reinforced Plastics and Composites, 28, 257.

5.    Kanga, S., Honga, S.I., Choeb, C.R., Parkb, M., Rimb, S. and Kimb, J. (2001) Preparation and Characterization of Epoxy Composites Filled with Functionalized Nanosilica Particles Obtained via Solgel Process. Polymer, 42, 879-887. http://dx.doi.org/10.1016/S0032-3861(00)00392-X

6.    Luyt, A.S., Molefi, J.A. and Krump, H. (2005) Thermal Mechanical and Electrical Properties of Copper Filled Low Density Ald Linear Low Density Polyethylene Composites. Polymer Degradation and Stability, 1-8.

7.    Yung, K.C., Wang, J. and Yue, T.M. (2008) Thermal Management for Boron Nitride Filled Metal Core Printed Circuit Board. Journal of Composite Materials, 42, 2615.http://dx.doi.org/10.1177/0021998308096326

8.    Zhou, W.Y., Wang, C.F., An, Q.L. and Ou, H.Y. (2008) On Thermal Properties of Heat Conductive Silicone Rubber Filled with Hybrid Fillers. Journal of Composite Materials, 42, 173.

9.    Mamunya, Ye.P., Davydenko, V.V., Piss, P. and Lebedev, E.V. (2002) Electrical and Thermal Conductivity of Polymers Filled with Metal Powders. European Polymer Journal, 38, 1887-1897. http://dx.doi.org/10.1016/S0014-3057(02)00064-2

10.    Zhou1, T., Wang, X., Cheng, P., Wang, T., Xiong, D. and Wang, X. (2013) Improving the Thermal Conductivity of Epoxy Resin by the Addition of a Mixture of Graphite Nanoplatelets and Silicon Carbide Microparticles. Express Polymer Letters, 7, 585-594. http://dx.doi.org/10.3144/expresspolymlett.2013.56

11.    Reine, B., Tomaso, J.D., Dusserre, G. and Olivier, P.A. (2012) Study of Thermal Behaviour of Thermoset Polymer Matrix Filled with Micro and Nano Particles. Proceedings of ECCM15, Italy, 24-28 June 2012.

12.    Pal, R. (2007) New Models for Thermal Conductivity of Particulate Composites. Journal of Reinforced Plastics and Composites, 26, 643.http://dx.doi.org/10.1177/0731684407075569

13.  Hong, J.-P., Yoon, S.-W., Hwang, T.-S., Lee, Y.-K., Won, S.-H. and Nam, J.-D. (2010) Inter Phase Control of Boron Nitride/Epoxy Composites for High Thermal Conductivity. Korea-Australia Rheology Journal, 22, 259-264.

# Chapter 7

# Strength Characterization of E-glass Fiber Reinforced Epoxy Composites with Filler Materials

K. Devendra[1] and T. Rangaswamy[2]

[1]Department of Mechanical Engineering, SKSVMACET, Laxmeshwar, India

[2]Department of Mechanical Engineering, Government Engineering College (GEC), Hassan, India

## ABSTRACT

In this research work, an investigation was made on the mechanical properties of E-glass fiber reinforced epoxy composites filled by various filler materials. Composites filled with varying concentrations of fly ash, aluminum oxide ($Al_2O_3$), magnesium hydroxide ($Mg(OH)_2$) and hematite powder were fabricated by standard method and the mechanical properties such as ultimate tensile strength, impact strength and hardness of the fabricated composites were studied. The test results show that composites filled by 10% volume $Mg(OH)_2$ exhibited maximum ultimate tensile strength and hardness. Fly ash filled composites exhibited maximum impact strength.

# INTRODUCTION

Polymers have replaced many of the conventional metals/materials in various applications. This is possible be- cause of the advantages such as ease of processing, productivity, cost reduction, etc. offered by polymers over conventional materials. In most of these applications, the properties of polymers are modified by using fibers to suit the high strength/high modulus requirements. The high performance of continuous fiber (e.g. carbon fiber, glass fiber) reinforced polymer matrix composites is well known and documented [1]. Among the thermosetting polymers, epoxy resins are the most widely used for high-performance applications, such as matrices for fiber reinforced composites, coatings, structural adhesives and other engineering applications. Epoxy resins are characterized by excellent mechanical and thermal properties, high chemical and corrosion resistance, low shrinkage on curing and the ability to be processed under a variety of conditions [2]. However, these composites have some disadvantages related to the matrix dominated properties which often limit their wide applications. In the industry, the addition of filler materials to a polymer is a common practice. This improves not only stiffness, toughness, hardness, heat distortion temperature, and mold shrink- age, but also reduces the processing cost significantly. In fact, more than 50% of all produced polymers are in one way or another filled with inorganic fillers to achieve the desired properties [3]. Mechanical properties of fiber- reinforced composites are depending on the properties of the constituent materials (type, quantity, fiber distribution and orientation, void content). Beside those proper- ties, the nature of the interfacial bonds and the mechanisms of load transfer at the interphase also play an important role [4]. Nowadays specific fillers/additives are added to enhance and modify the quality of composites as these are found to play a major role in determining the physical properties and mechanical behavior of the composites. For many industrial applications of glass fiber reinforced epoxy composite, information about their mechanical behavior is of great importance. Therefore, in this work, the mechanical behavior of E-glass fiber reinforced epoxy composites filled by varying concentration of fly ash $Al_2O_3$, Mg $(OH)_2$ and hematite powder has been studied.

# EXPERIMENTATION

## Materials for Composites

ARALDITE (L-12) epoxy had been used as matrix material for reinforced composites in this experimental work. For reinforcing epoxy matrix, E-glass fiber was used along with hardener K-6. Fly ash, $Al_2O_3$, $Mg(OH)_2$ and hematite powder was used as filler materials. Fly ash was obtained from thermal power plant. Measured density of fly ash particles was 2.0 g/cc. These micron-sized elements are consists primarily of silica (59.5%), alumina (20.3%), $Fe2O3$ (6.5%), titanium (1.28%), potassium ox- ide (0.96%), MgO (0.50%), Phosphates (0.05%), Sulfates (0.345%) and unburned coal [5]. Most of these particles have a gas bubble at the center. Aluminum oxide particles is a ceramic powder commonly used filler, it is also used as an abrasive due to its hardness. Magnesium hydroxide is an inorganic compound and it is a white powder with specific gravity of 2.36, very slightly soluble in water; decomposing at 350°C. Magnesium hydroxide is attracting attention because of its performance, price, low corrosiveness and low toxicity. Hematite is an iron oxide with the same crystal structure as that of corundum (rubies and sapphires). Usually the color varies between a metallic grey and red. Hematite is one of the oldest stones mined in our history.

## Fabrication of Composites

The E-glass/Epoxy based composite slabs filled with varying concentrations of (0%, 10% and 15% volume) fly ash, aluminum oxide $(Al_2O_3)$, magnesium hydroxide $(Mg(OH)_2)$, and hematite powder were prepared. The volume fraction of fiber, epoxy and filler materials were determined by considering the density, specific gravity and mass. Fabrication of the composites is done at room temperature by hand layup techniques. The required ingredients of resin, hardener, and fillers are mixed thoroughly in a basin and the mixture is subsequently stirred constantly. The glass fiber positioned manually in the open mold. Mixture so made is brushed uniformly, over the glass plies. Entrapped air is removed manually with squeezes or rollers to complete the laminates structure and the composite is cured at room temperature.

## Specimen Preparation

The prepared E-glass fiber reinforced epoxy composite slabs filled by various filler materials were taken out from the mold and then specimens of suitable dimensions were prepared from the composite slabs for different mechanical tests according to ASTM standards. The test specimens were cut by slabs by using diamond tipped cutter and different tools in the work shop. Three identical test specimens were prepared for different test. Designation and composition of prepared composite slabs are presented in Table 1.

# MECHANICAL PROPERTY TESTING

Mechanical properties of composites were evaluated by tensile, impact and hardness measurements. Tensile, impact and hardness tests were carried out using Universal testing machine, impact machine and hardness testing machine respectively. Three identical samples were tested for tensile strength, impact strength and hardness.

**Table 1:** Designation and compositions of fabricated com- posites

| Material Designation | Glass Fiber (Volume %) | Epoxy (Volume %) | Filler Materials (Volume %) |
|---|---|---|---|
| GE | 50 | 50 | Nil |
| GEF$_1$ | 50 | 40 | 10% Fly ash |
| GEF$_2$ | 50 | 35 | 15% Fly ash |
| GEA$_1$ | 50 | 40 | 10% Al2O3 |
| GEA$_2$ | 50 | 35 | 15% Al$_2$O$_3$ |
| GEM$_1$ | 50 | 40 | 10% Mg(OH)$_2$ |

| GEM$_2$ | 50 | 35 | 15% Mg(OH)$_2$ |
| GEH$_1$ | 50 | 40 | 10% Hematite Powder |
| GEH$_2$ | 50 | 35 | 15% Hematite Powder |

# Ultimate Tensile Strength

Tensile tests were examined according to ASTM D3039 using a universal testing machine at room temperature. Test specimens having dimension of length 250 mm, width of 25 mm and thickness of 2.5 mm. The specimen was loaded between two manually adjustable grips of a 60 KN computerized universal testing machine (UTM) with an electronic extensometer. Test was repeated thrice and the average value was taken to calculate the tensile strength of the composites.

## *Details of Universal Testing Machine*

Universal testing machine is a Micro Control Systems make and model MCS-UTE60 and software used is MCSUTE STDW2KXP. System uses add-on cards for data acquisition with high precision and fast analog to digital converter for pressure/Load cell processing and rotary encoder with 0.1 or 0.01 mm for measuring cross head displacement (RAM stroke). These cards are fitted on to slots provided on PC's motherboard WINDOW9X based software is designed to fulfill nearly all the testing requirements. MCS make electronic extensometer is used with an extremely accurate strain sensor for measuring the strain of the tensile samples.

# Impact Strength

The Charpy impact strength was carried out on composites in accordance with ASTM E23 using impact testing machine. The dimensions of the specimens were 10 mm × 10 mm × 55 mm size on one side surface of the specimen a V-notch have been made at an angle of 45° with root depth of 2 mm. Test was repeated thrice and the average values were taken for calculating the impact strength.

## Brinell Hardness Test

Brinell hardness test was conducted on the specimen using a standard Brinell hardness tester. A load of 250 kg was applied on the specimen for 30 sec using 5 mm diameter hard metal ball indenter and the indentation diameter was measured using a microscope. The hardness was measured at three different locations of the specimen and the average value was calculated. The indentation was measured and hardness was calculated using Equation (1).

$$BHN = \frac{2P}{\pi D(D - \sqrt{(D^2 - d^2)})}$$

(1)

where: $P$ = Applied force (Kgf); $D$ = Diameter of indenter (mm); $d$ = Diameter of indentation (mm).

# RESULTS AND DISCUSSION

Results obtained from this experimental work are presented in Tables 2-4 and Figures 1-3. Mechanical properties of fiber-reinforced epoxy composites are depending on the properties of the constituent materials (type, quantity, fiber distribution and orientation, void content). Beside those properties, the nature of the interfacial bonds and the mechanisms of load transfer at the inter phase also play an important role.

## Ultimate Tensile Strength

The tensile strength of the E-glass fiber reinforced epoxy composites depends upon the strength and modulus of the fibers, strength and chemical stability of the matrix, fiber matrix interaction and fiber length.

**Table 2:** Comparison of ultimate tensile strength

| Composite materials | Ultimate Tensile Strength, (MPa) |
|---------------------|----------------------------------|
| GE | 450.24 |
| GEF$_1$ | 249.80 |

| | |
|---|---|
| GEF$_2$ | 168.80 |
| GEA$_1$ | 297.80 |
| GEA$_2$ | 257.21 |
| GEM$_1$ | 375.36 |
| GEM$_2$ | 347.20 |
| GEH$_1$ | 156.92 |
| GEH$_2$ | 182.30 |

**Table 3:** Comparison of Charpy impact strength

| Composite materials | Charpy Impact Strength (J/mm$^2$) |
|---|---|
| GE | 0.2846 |
| GEF$_1$ | 0.2041 |
| GEF$_2$ | 0.1500 |
| GEA$_1$ | 0.1681 |
| GEA$_2$ | 0.1575 |
| GEM$_1$ | 0.1687 |
| GEM$_2$ | 0.1625 |
| GEH$_1$ | 0.1250 |
| GEH$_2$ | 0.1583 |

**Table 4:** Comparison of Brinell hardness number

| Composite materials | Hardness (BHN) |
|---|---|
| GE | 57.64 |
| GEF$_1$ | 39.68 |
| GEF$_2$ | 37.11 |
| GEA$_1$ | 73.90 |
| GEA$_2$ | 82.13 |
| GEM$_1$ | 88.69 |
| GEM$_2$ | 88.10 |
| GEH$_1$ | 46.86 |
| GEH$_2$ | 54.25 |

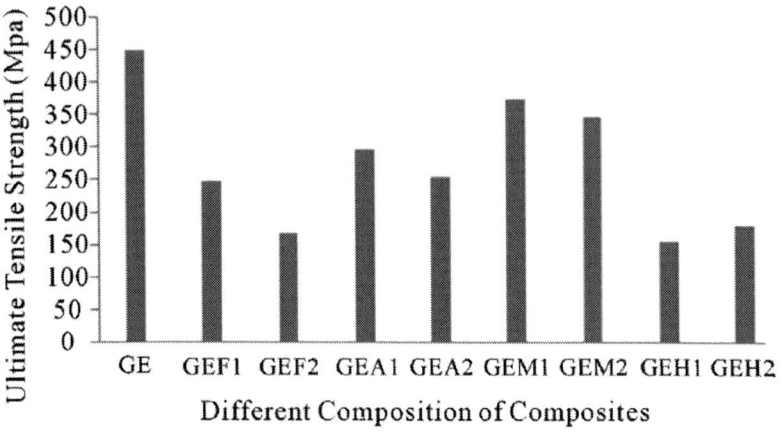

**Figure 1:** Ultimate tensile strength for different composition of composite materials.

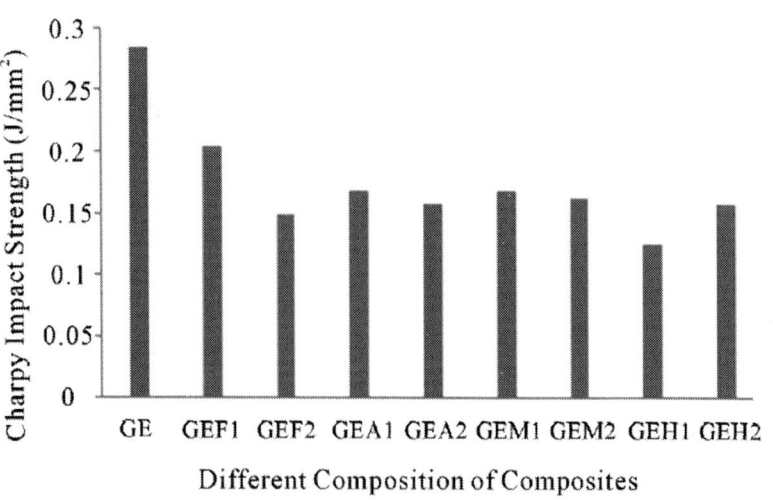

**Figure 2:** Charpy impact strength for different composition of composite materials.

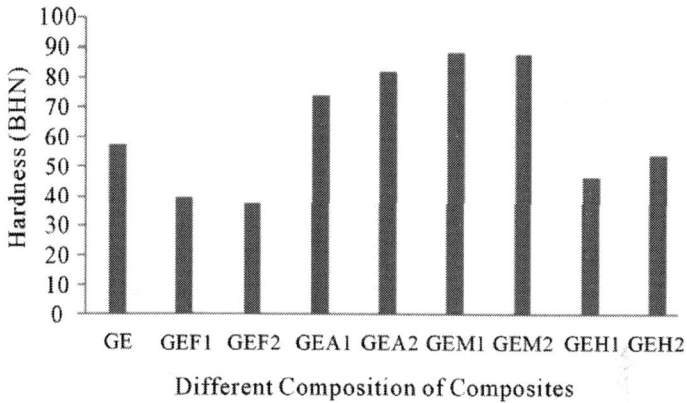

**Figure 3:** Brinell hardness number for different composition of composite material.

From the obtained results it was observed that composite filled by 10% Volume $Mg(OH)_2$ exhibited maximum ultimate strength of 375.36 MPa when compared with other filled composites but lower than the un filled composite [Figure 1]. This may be due to good particle dispersion and strong polymer/filler interface adhesion for effective stress transfer. Composites filled by $Al_2O_3$ exhibited better ultimate tensile strength compared with composites filled by fly ash and hematite this is due to that $Al_2O_3$ having the ceramic particles these particles distributed uniformly throughout the composites and produces good bonding strength between polymer, filler and fiber. But increase in addition of $Mg(OH)_2$, $Al_2O_3$ and fly ash content up to 15% volume to the composites the tensile strengths is found to be less this is due to more filler material in the composites damages matrix continuity, less volume of fiber and more void formation in the composites. Ultimate tensile strength increases with increase in addition of hematite to composites this may be due to improved in inter facial bonding strength between filler, matrix and fiber.

# Impact Strength

Impact strength is defined as the ability of a material to resist the fracture under stress applied at high speed. The impact properties of composite materials are directly related to overall toughness and composite fracture toughness is affected by inter laminar and interfacial strength parameters.

From Figure 2, it is observed that composite filled by 10% volume fly ash having high impact strength when compared with other filled composites this is due to that good bonding strength between filler, matrix, fiber and flexibility of the interface molecular chain resulting in absorbs and disperses the more energy, and prevents the cracks initiator effectively. But there was reduction in impact resistance as the fly ash content increases which might be because of formation of additional voids and this void increases the crack propagation. Impact strength decreases when increase in addition of $Al_2O_3$ and $Mg(OH)_2$ to composites. Typically, a polymer matrix with high loading of fillers has less ability to absorb impact energy this is because the fillers disturb matrix continuity and each fillers is a site of stress concentration, which can act as a micro crack initiator and reduces the adhesion and energy absorption capacity of composites. Test results show that impact strength increases with adding more hematite powder to composites this due to improvement of bonding strength between filler and matrix and rigidity of filler particles absorbs the more energy.

# Hardness

Hardness properties of all the composites are presented in the Table 4.

The experimental results show that composite filled by 10% volume $Mg(OH)_2$ exhibited maximum hardness number of 88.69 BHN when compared with other filled composites this due to uniform dispersion of $Mg(OH)_2$ particles and good bonding strength between fiber and matrix [6]. From Figure 3, it is observed that increase in addition of $Al_2O_3$ and hematite to composites leads to increase in hardness number this may be due to the improved bond between the matrix and reinforcement, reduced porosity. When increasing the particle loading in the matrix decreases the inter particle distance with results in increase of resistance to indentation. Fly ash filled composites exhibited less hardness number this due to weak bonding strength and more possibility of void formation.

# CONCLUSIONS

Based upon the test results obtained from the various tests carried out, following conclusions were made:

- From the obtained results, it was observed that composite filled by 10% volume of Mg $(OH)_2$ exhibited maximum ultimate strength of 375.36 MPa when compared with other filled composites. Composites filled by Al2O3 exhibited better ultimate strength compared with composites filled by fly ash and hematite. Increase in addition of Mg $(OH)_2$, $Al_2O_3$ and fly ash to composites leads to decrease in ultimate tensile strength.

- Experimental results show that composites were filled by 10% volume of fly ash having high impact strength when compared with other filled composites. Composites filled by 10% volume $Al_2O_3$ and Mg$(OH)_2$ exhibited good impact strength but increase in addition of $Al_2O_3$ and Mg$(OH)_2$ leads to decrease in impact strength. Test results indicated that impact strength increases with adding more hematite powder to composites.

- The experimental results indicated that composite filled by Mg $(OH)_2$ exhibited maximum hardness number 88.69 BHN when compared with other filled composites. From the results, it is observed that increase in addition of $Al_2O_3$ and hematite to composites increases the hard- ness of the composites. Increase in addition of fly ash to composites leads to decrease in hardness number.

# REFERENCES

1.  Yasmin and I. M. Daniel, "Mechanical and Thermal Properties of Graphite Platelet/Epoxy Composites," *Polymer*, Vol. 45, No. 24, 2004, pp. 8211-8219. http://dx.doi.org/10.1016/j. polymer.2004.09.054

2.  N. Hameed, P. A. Sreekumar, B. Francis, W. Yang and S. Thomas, "Morphology, Dynamic Mechanical and Thermal Studies on Poly(styrene-co-acrylonitrile) Modified Epoxy Resin/Glass Fibre Composites," *Composites Part* A, Vol. 38, No. 12, 2007, pp. 2422-2432. http://dx.doi.org/10.1016/j.compositesa.2007.08.009

3.    R. N. Rothon, "Mineral Fillers in Thermoplastics: Filler Manufacture and Characterization," *Advances in Polymer Science*, Vol. 139, 1999, pp. 67-107.

4.    Cs. Varga, N. Miskolczi, L. Bartha and G. Lipoczi, "Improving the Mechanical Properties of Glass-Fibre-Reinforced Polyester Composites by Modification of Fibre Surface," *Materials and Design*, Vol. 31, 2010, pp. 185-193.

5.    M. Single and V. Chawla, "Mechanical Properties of Epoxy Resin-Fly Ash Composite," *Journal of Minerals and Materials Characterization and Engineering*, Vol. 9, No. 3, 2010, pp. 199-210.

6.    R. K. Goyal, A. N. Tiwari and Y. S. Negi, "Micro Hardness of PEEK/Ceramic Micro- and Nano Composites: Correlation with Halpin-Tsai Model," *Materials Science and Engineering*, Vol. 491, No. 1-2, 2008, pp. 230-236.

# Chapter 8

# Adding Sn on the Performance of Amorphous Brazing Fillers Applied to Brazing TA2 and Q235

Jie Cui[1], Qiuya Zhai[1], Jinfeng Xu[1], Yahui Wang[2], and Jianlin Ye[2]

[1]Xi'an University of Technology, Xi'an, China
[2]Xi'an Unit Container Manufacturing LTD Co., Xi'an, China

## ABSTRACT

Amorphous filler Ti-Zr-Cu-Ni with better performance and higher melting point than the $\alpha \rightarrow \beta$ phase transition temperature 882.5°C of TA2, is appropriate for joining TC and TB titanium alloys but not for TA titanium alloys, with which the ductility of the joined base metal TA2 gets down. In this paper, Sn is added into Ti-Zr-Cu-Ni filler to reduce its melting temperature then to satisfy the joining temperature requirement, and the effects of the content of Sn on the microstructure of the alloy and brazing performance are investigated. The results show that, the Ti-Zr-Cu-Ni-Sn brazing foils still possess amorphous structure; the melting point of fillers is reducing with the increase of the Sn content; the joint gap that formed during brazing TA2 and Q235 using Ti-Zr-Cu-Ni-Sn foils is fully filled with continuous compact surface and smooth uniform fillet; the shear strength of the joint is raising with the increase of Sn content in brazing fillers and the strength reaches to 112 MPa when Sn content is 3%; adding more Sn, more brittle intermetallic compounds TiFe and $TiFe_2$ are gathering to form cluster and the shear strength of the joint is reducing; the shear fracture always occurs in the center of the seam.

# INTRODUCTION

Titanium alloys have high specific strength, low density, good thermo stability, tenacity, thermal conductivity and high fatigue resistance but high price. Common engineering material mild steel Q235 has excellent performance but reasonable price. Therefore, the advantages of these two materials above-mentioned can be made full use of by connecting, which has high practical values and good economic benefits.

It is hard to connect titanium and mild steel for the great difference in their physical and chemical properties. There are many methods to connect titanium and titanium alloy at present, such as argon arc welding, electron beam welding, laser welding and other fusion welding; diffusion welding, friction welding, explosion welding and other solid-state joining; infrared welding, high frequency welding, vacuum brazing and other brazing [1] - [4] . Among them, brazing has important application in aerospace field with low welding temperature, simple equipment and process, which is applied to the connection of dissimilar metals titanium and steel and is available for the weld of precision components with complex structure and thin-wall.

There is a great variety of brazing fillers for brazing titanium and titanium alloy, such as Ag-based, Al-based, Pb-based and Ti-based brazing fillers. Among them, Ti-based brazing filler can be made into amorphous filler through rapid solidification. This kind of filler has pure uniform ingredient, low welding temperature, good wettability, brazing quality and many other outstanding merits.

In 1990, Onzawa T. et al. developed amorphous brazing filler Ti-Zr-Cu-Ni which now has become an important filler for brazing titanium and titanium alloy, and the joint can work in high temperature and highly corrosive media [5] . However, this kind of filler is mostly applied to the connection of titanium and titanium alloy, rarely to the connection of dissimilar metals titanium and steel. The melting temperatures of this kind of fillers are almost in the range of 980°C - 1000°C much higher than the $\alpha \rightarrow \beta$ phase transition temperature of TA2. When the welding temperature is higher than 882.5°C, titanium translates from $\alpha$ phase with close-packed hexagonal structure to $\beta$ phase with body-centered cubic structure, and then translates into acicular structure in the subsequent process of rapid solidification, which reduces the ductility of the base metal titanium near the brazing seam. Accordingly, in order to avoid the phase transformation, the melting temperature of the filler should be lowered and the brazing temperature should be controlled under the phase transformation temperature.

Adding element that reduces the brazing temperature is a good solution to this problem. Beryllium can form limit solid solution and compound with titanium and small content of it can reduce the melting temperature. Other elements such as V, Cr, Fe and Co can also have the same effect but they are not obvious [6]. Sn without poison or harm can not only reduce the melting temperature but also barely change alloy's phase composition with small content [7].

In this paper, in order to obtain an amorphous filler with excellent performance that can be applied to the brazing of TA2/Q235, amorphous filler Ti-Zr-Cu-Ni-Sn is designed; filler foils are prepared by using a single roller rapid solidification apparatus; fillers' properties are tested; the brazing of TA2/Q235 with the designed fillers is investigated; the effect of the content of Sn on microstructure and mechanical properties of the joint are analyzed.

# DESIGN OF EXPERIMENTS

In this experiment, element Sn that can lower the melting temperature is added into the master alloy $Ti_{35}Zr_{35}Cu_{15}Ni_{15}$. The compositions of six designed fillers are showed in Table 1.

**Table 1**: Chemical compositions of designed brazing fillers Ti-Zr-Cu-Ni-Sn

| Num. | Elements (wt%) | | | | |
|------|------|------|------|------|------|
|      | Ti   | Zr   | Cu   | Ni   | Sn   |
| 1    | 35   | 35   | 35   | 35   | 0    |
| 2    | 34.65| 34.65| 14.85| 14.85| 1    |
| 3    | 33.95| 33.95| 14.55| 14.55| 3    |
| 4    | 33.25| 33.25| 14.25| 14.25| 5    |
| 5    | 32.55| 32.55| 13.95| 13.95| 7    |
| 6    | 31.85| 31.85| 13.65| 13.65| 9    |

Simple metals (99.99%) were smelted into brazing filler metal by high frequency induction heating equipment in argon atmosphere and brazing fillers were prepared by using a single roller rapid solidification apparatus [8] - [10] in this experiment. The experimental parameters can be seen in reference [9]. The brazing fillers with thickness of 40 µm - 60 µm and width of 4 mm, the biggest length just 30 cm, and their phase structure and composition were tested by X-ray diffractometer, were studied by means of different scanning calorimeter to analyze the effect of the content of Sn on the melting temperature of brazing fillers.

The brazing is conducted in vacuum high frequency brazier, the brazing parameters are: vacuum degree is 0.1 Pa; welding temperature is 800°C; heating current is 25 A; heating time is 15 s; holding time is 15 s; cooling to room temperature is in furnace. The sample made along the axis of the welded sample is etched with the solution of 3 ml HF + 6 ml $HNO_3$ + 100 ml $H_2O$ [11], the effect of the content of Sn on the microstructure of the joint is analyzed by using Olympus GX-71 metallurgical microscope, the shear strength of the joints is tested by WE-100 universal testing machine and the effect of the content of Sn on the strength is analyzed.

# RESULTS AND DISCUSSION

## The Effects of Adding Sn on the Microstructure and Properties of the Brazing Fillers

Figure 1 shows the X-ray diffraction spectrum patterns of Ti-Zr-Cu-Ni-Sn brazing fillers. In the pictures there is no peak according to crystal phase, but broad diffraction peaks belong to glassy phases only. The brazing fillers still process typical amorphous diffraction peaks when the content of Sn rises to 7%. Besides the corresponding base peaks of the fillers with 7% Sn, some peaks get sharp in local XRD curve of the brazing filler when the content reaches to 9%. It can thus be seen that the amorphous structure of the brazing fillers cannot be changed by adding a bit Sn, but tend to be crystallized when the content increases rapidly.

Figure 2 shows the DSC curves of the three brazing filler metals. In Figure 2(a), it can be seen that the melting temperature of brazing filler $Ti_{33.95}Zr_{33.95}Cu_{14.55}Ni_{14.55}Sn_3$ is the highest of the three, almost near the $\alpha \rightarrow \beta$ phase transition temperature 882.5°C of TA2.

Brazing filler $Ti_{33.25}Zr_{33.25}Cu_{14.25}Ni_{14.25}Sn_5$ begins to melt at 843°C and stops melting at 862°C in Figure 2(b). The melting temperature interval of 843°C - 862°C, which is small, nearly meets the using demand for brazing $\alpha$-Ti and steel.

It can be seen from Figure 2(c) that brazing filler $Ti_{31.85}Zr_{31.85}Cu_{13.65}Ni_{13.65}Sn_9$ begins to melt at 596°C and stops melting at 615°C. The small melting temperature interval of 596°C - 615°C, meets the using requirement for brazing $\alpha$-Ti and steel. With an increase in content of Sn, the melting temperature is dropping. The melting point of $Ti_{32.55}Zr_{32.55}Cu_{13.95}Ni_{13.95}Sn_7$, between the ones of $Ti_{33.25}Zr_{33.25}Cu_{14.25}Ni_{14.25}Sn_5$ and $Ti_{31.85}Zr_{31.85}Cu_{13.65}Ni_{13.65}Sn_9$, meets the using requirement for brazing $\alpha$-Ti and steel as well.

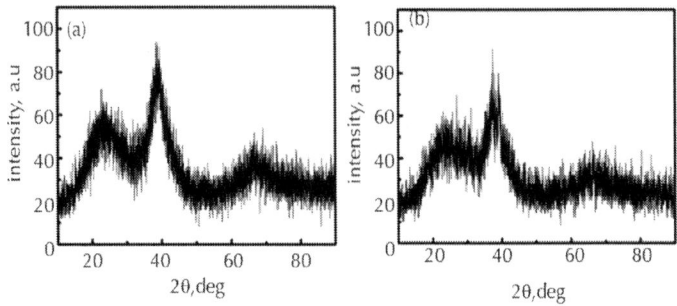

**Figure 1**: X-ray diffraction pattern of brazing filler metal. (a) x = 7; (b) x = 9.

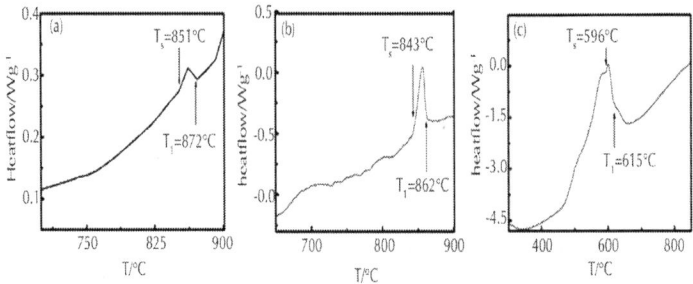

**Figure 2**: DSC curves of brazing ribbon of $(Ti_{35}Zr_{35}Cu_{15}Ni_{15})_{100-x}Sn_x$ (a) x = 3; (b) x = 5; (c) x = 9.

These brazing fillers are all viscous liquid under the welding temperature, their mobility can be measured by the viscosity of liquid metal. The bigger the viscosity is, the smaller the mobility becomes. And the viscosity is inversely proportional to the degree of superheat. Hence, with the increase of the content of Sn, the melting temperature of the alloy gets lower and the degree of superheat of the liquid metal becomes bigger, which will makes the viscosity lower, the mobility higher and then improves the spreadability of the brazing fillers under a certain welding temperature. Therefore, it can achieve the goal of reducing the melting temperature of the brazing fillers by adding Sn into the master alloy $Ti_{35}Zr_{35}Cu_{15}Ni_{15}$. And adding Sn can reduce the solidus and liquidus, shrink the melting temperature interval, improve the mobility, spreadability and welding performance of the brazing fillers [12] [13] .

# The Effects of Sn Content on the Microstructure and Properties of the TA2/Q235 Joints

The microstructure of the overlap joint TA2/Ti$_{31.85}$Zr$_{31.85}$Cu$_{13.65}$Ni$_{13.65}$Sn$_9$/Q235 is showed inFigure 3. Figure 3(a) shows the whole morphology of the joint. There appears to have three zones from top to bottom: base metal TA2 zone, brazing seam zone and base metal Q235 zone. The base metal TA2 near the brazing seam presents saw teeth shape for it partly converts into lath-like structure of β phase. However, the base metal TA2 away from the brazing seam remains the original structure of α phase.

It can divided into three parts of the brazing seam: firstly, the wider pale zone of the seam near the side of the base metal TA2 is the transition zone of the seam and the base metal TA2, containing a lot of Ti and fewer C and Fe. Secondly, the center of the seam consists of white sheet phase and gray matrix, and the small size white phases near the center gather to form cluster; the white phases that are brittle intermetallic compounds with lager size are sparsely distributed. Comparing with that, the transition region between the seam and the base metal Q235 appears to become narrower where there is an obvious boundary.

Some points (A, B, C, D) in the seam were examined by means of SEM showed in Figure 3(b). The white sheet phases mainly consisting of more Ti, Fe and small content of Ni, Cu and Zr are TiFe and TiFe$_2$. For element Fe can strongly propel β-Ti to remain at room temperature, these phases should be β-Ti solid solution including alloys. The matrix is mainly constituted by element Ti and lager content of C and Fe coming from the base metal Q235 which illustrates that the elements of the base metal Q235 have been well distributed into the seam.

Figure 4 shows the microstructure in the brazing seam zone of the TA2/Q235 joint with the brazing fillers including different contents of Sn. The microstructure of the TA2/Q235 joint without Sn is shown in Figure 4(a) and there appears to be some white phases distributed sparsely in the light color matrix. Figure 4(b) shows the microstructure of the TA2/Q235 joint with 5% Sn. This seam is constituted by light and dark color matrixes and there are white phases densely distributed in the light matrix with smaller size than the white phases in Figure 4(a).

The microstructure of the TA2/Q235 joint including 7% Sn is shown in Figure 4(c). There are white phases with significant difference in size and the larger ones are sparsely distributed, the smaller ones gather to form cluster. Figure 4(d) shows the microstructure of the TA2/Q235 joint with 9% Sn. Through comparing the microstructure of the TA2/Q235 joint with the brazing fillers including different content of Sn, it can be seen that, with an increase in Sn content, the microstructures of the seams change from uniformity to layering gradually, the size of the distributed white brittle phases in the seam is being diminished, so that the more and more white phases are gathering to form cluster.

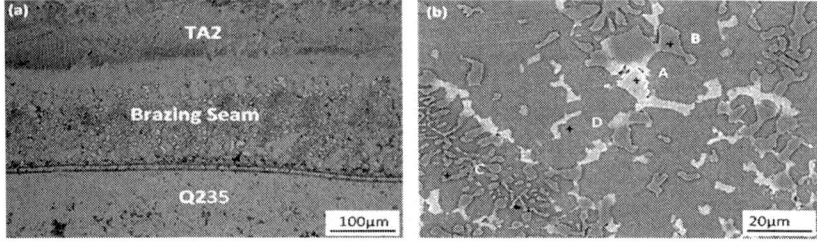

**Figure 3**: Microstructure of joint brazed with $Ti_{31.85}Zr_{31.85}Cu_{13.65}Ni_{13.65}Sn_9$. (a) Microstructure of joint; (b) Microstructure of seam.

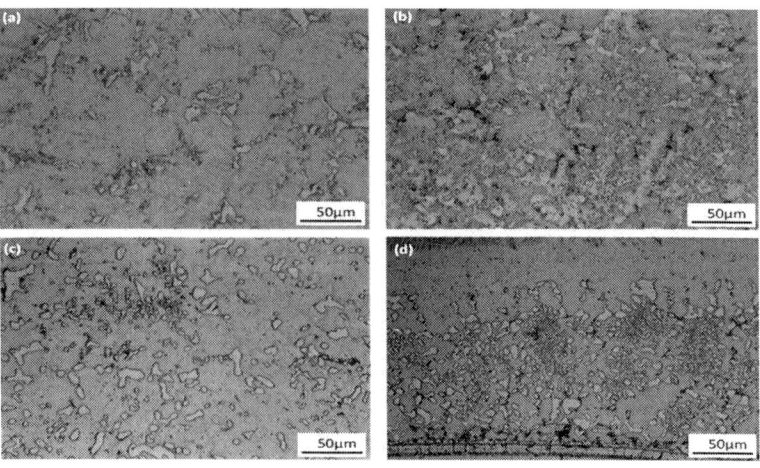

**Figure 4**: Microstructure of joint brazed with $(Ti_{35}Zr_{35}Cu_{15}Ni_{15})_{100-x}Sn_x$. (a) x = 0; (b) x = 5; (c) x = 7; (d) x = 9.

The solid solubility of Fe in α-Ti is just 0.05% - 0.1% at room temperature, not exceeding 0.5% at eutectoid temperature. Adding more Sn can reduces the melting temperature and the dissolution of Fe in Ti as well. When the content of Fe in β-Ti reaches to a certain amount, during the cooling process, the oversaturation of Fe in Ti occurs. When the content of Fe exceeds its solid solubility in Ti, the brittle intermetallic $TiFe_2$ and TiFe are formed. And obviously the appearance and augment of these white brittle phases reduces the mechanical properties of the joints.

## The Effect of the Content of Sn on the Mechanical Properties of the TA2/Q235 Joints

Figure 5 shows the shear strength curve of TA2/Q235 joints with the change of the amount of Sn. Through testing the shear strength of these TA2/Q235 joints with different contents of Sn, it is found that with an increase in the content of Sn, the shear strengths of TA2/Q235 joints firstly increase, reaching to 112 MPa, and then descend. From the appearance of the facture, it can be seen that the cracks mainly occur in the center of the seam. This is because more brittle intermetallic compounds are emerged in the center of the seam with too much more Sn than the seam without Sn, which descends the strength of the joints. And the ductility of the base metal TA2 of the joint kept by adding Sn to decrease the α → β phase transition cannot compensate the harm to the property of the joints from the emerged intermetallic compounds. When the content of Sn is 3%, the α → β phase transition and the intermetallic compounds are both less, which enhances the strength of the joints.

# CONCLUSIONS

- In this paper, in order to reduce the melting temperature of the Ti-Zr-Cu-Ni master alloy, three brazing fillers $(Ti_{35}Zr_{35}Cu_{15}Ni_{15})_{100-x}Sn_x$ (x = 5, 7, 9) with amorphous structure are designed and prepared. With an increase in the content of Sn, the shear strength and the melting temperature are gradually descending, satisfying the welding temperature requirement for brazing TA2/Q235.

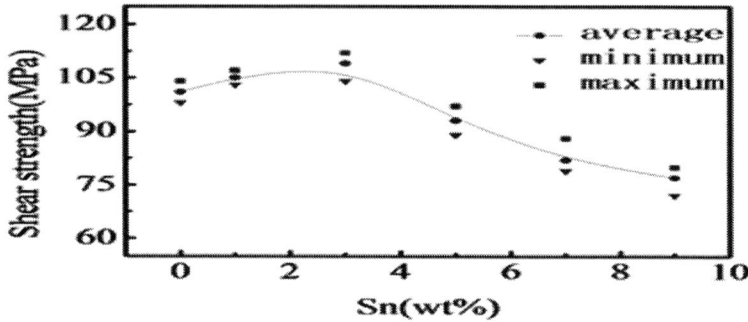

**Figure 5**: Influence of Sn contents on shear strength of TA2/Q235 joints.

- The high frequency induction vacuum brazing of TA2/Q235 is conducted by using designed amorphous fillers with lower melting point, which has good performance. With an increase in the content of Sn, white brittle intermetallic compounds TiFe and TiFe$_2$ with reducing size increase little by little gathering to form cluster; the shear strengths of the joints firstly increase, reaching to 112 MPa when the Sn content is 3%, and then descend. The cracks mainly occur in the center of the seams.

# REFERENCES

1. Meshra, S.D. and Mohandas, T. (2010) A Comparative Evaluation of Friction and Electron Beam Welds of Near-α Titanium Alloy. Materials and Design, 31, 2245-2252. http://dx.doi.org/10.1016/j.matdes.2009.10.012

2. Liu, J., Watanabe, I. and Yoshida, K. (2002) Joint Strength of Laser-Welded Titanium. Dental Materials, 18, 143-148. http://dx.doi.org/10.1016/S0109-5641(01)00033-1

3. Chang, C.T., Du, Y.C. and Shiue, R.K. (2006) Infrared Brazing of High-Strength Titanium Alloys by Ti-15Cu-15Ni and Ti-15Cu-25Ni Filler Foils. Materials Science and Engineering, 420, 155-164. http://dx.doi.org/10.1016/j.msea.2006.01.046

4. Elrefaey, A. and Tillmann, W. (2007) Interface Characteristics and Mechanical Properties of the Vacuum-Brazed Joint of Titanium-Steel Having a Silver-Based Brazing Alloy. Metallurgical and

Materials Transactions, 38, 2956-2961.http://dx.doi.org/10.1007/s11661-007-9357-5

5.  Shapiro, A.E. and Flom, Y.A. (2012) Brazing of Titanium at Temperature below 800°C: Review and Prospective Applications. Welding Journal, 50, 1-22.

6.  Chang, C.T., Shiue, R.K. and Chang, C.S. (2005) Microstructural Evolution of Infrared Brazed Ti-15-3 Alloy Using Ti-15Cu-15Ni and Ti-15Cu-25Ni Fillers. Scripta Materialia, 54, 853-858. http://dx.doi.org/10.1016/j.scriptamat.2005.11.013

7.  Qi, Y., Zhang, Y.H. and Quan, B.Y. (2003) Development and Application of Braze Welding and Ti-Based Braze Material. Metallic Functional Materials, 10, 31-37.

8.  Huang, Y.J., et al. (2008) Formation, Thermal Stability and Mechanical Properties of $Ti_{42.5}Zr_{7.5}Cu_{40}Ni_5Sn_5$ Bulk Metallic Glass. Science in China Series G: Physics, Mechanics Astronomy, 51, 372-378. http://dx.doi.org/10.1007/s11433-008-0049-y

9.  Xu, J.F. and Wei, B.B. (2004) Liquid Phase Flow and Microstructure Formation during Rapid Solidification. Acta Physica Sinica, 53, 160-166.

10. Threadgill, P.L. and Dance, B.C.I. (1971) Joint of Intermetallic Alloys Further Studies. Welding Journal, 50, 379.

11. Onzawa, T., Suzumura, A. and Ko, M. (2012) Structure and Mechanical Properties of CP Ti and Ti-6Al-4V Alloy Joints Brazed with Ti-Based Amorphous Filler Metals. Japan Welding Society, 5, 205-211.

12. Botstein, O. and Rabinkin, A. (1994) Brazing of Titanium-Based Alloys with Amorphous Ti-25Zr-50Cu Filler Metal. Materials Science and Engineering A, 188, 305-315.http://dx.doi.org/10.1016/0921-5093(94)90386-7

13. Howden, D.G. and Monroe, R.W. (1972) Suitable Alloys for Brazing Titanium Heat Exchangers. Welding Journal, 51, 31-36.

# Chapter 9

# Cellulose Microfibril from Banana Peels as a Nanoreinforcing Fillers for Zein Films

Manisara Phiriyawirut, Parichat Maniaw

Department of Tool and Materials Engineering, Faculty of Engineering, King Mongkut's University of Technology Thonburi, Bangkok, Thailand

## ABSTRACT

Cellulose microfibril (CMF) was the extraction with acid mixture from peel of Musa sapientum Linn type of banana (Kluai Nam Wa). The fibrous-shape of CMF interconnected weblike structure with the average diameter 26 nm were observed by TEM. In order to prepare zein/CMF nanocomposite films, 16 wt% zein solution was prepared by dissolved in 80% ethanol aqueous solution which contain glycerol 20% w/w. The suspension of CMF and zein solution was mixed with 0% - 5% weight fractions of solid CMF in zein matrix. The morphology of the zein films is more roughness by increased amount of cellulose microfibrils. It was found that as CMF content increase from 0 to 5 wt% results in increasing tensile strength and Young's modulus of zein nanocomposite films. The highest strength obtains at 4 wt% CMF.

# INTRODUCTION

Natural fibers are subdivided based on their origins, coming from plants, animals or minerals. Especially, natural fibers which come from plant, they are also referred to as cellulosic fibers, related to the main chemical component cellulose, or as lignocellulosic fibers, since the fibers usually often also contain a natural polyphenolic polymer, lignin, in their structure. In recent years, the use of natural fibers as reinforcements in polymers and composites has attracted much attention due to the environmental concerns such as sisal [1], cotton [2], bamboo [3], jute [4], kenaf [5] and wood [6]. Compared to inorganic fillers, the main advantages of lignocellulosics are renewable nature, wide variety of fillers available throughhout the world, low energy consumption, low cost, low density, high specific strength and modulus (desirable fiber aspect ratio), high sound attenuation, and comparatively easy processability due to their flexibility and nonabrasive nature which allow high filling levels.

Nevertheless, despite these attractive properties, lignocellulosic fillers are used only to a limited extent in industrial practices, mainly due to difficulties associated with surface interactions. An alternative way to palliate this restriction consists of obtaining both components (matrix and filler) dispersed in water such as colloidal suspensions of cellulose whiskers. In addition, cellulose can also be used as a microfibrillar filler, which is more accessible in terms of available amounts and preparation. Cellulose microfibrils (CMF) can be found as intertwined microfibrils in the parenchyma cell wall, in particular from plant.

Cellulose whiskers have generated a great deal of interest as a source of nanometer size filler because of very good mechanical properties [7]. The microfibril diameters range from 2 to 20 nm and their lengths can reach several tens of microns depending on their origin, thus they have a very high aspect ratio and a significant load-carrying capability. Another type of cellulose, CMF is a diameter range of 10 - 100 nm, but with a web-like structure [8]. Cellulose nanowhiskers and CMF have been isolated from various sources such as sugar beet [9], potato pulps [10], algae [11], cactuses [12], banana rachis [13] and banana peel [14].

The use of cellulose nanowhiskers and CMF in composite materials has been reviewed [7]. Both natural and synthetic polymers have been used as matrices including starch [15], chitosan [16], poly(vinyl chloride) [17], and natural rubber [14], etc. Most of the composites have shown a significant improvement in properties.

In this work, CMF which is prepared from peel of Musa sapientum Linn type of banana (Kluai Nam Wa) was used as reinforcement filler of zein films which prepared by solution casting technique. The effect of CMF content on morphology, chemical structure, crystal structure, thermal and mechanical properties of composite films were investigated and reported here.

# EXPERIMENTAL

## Materials

CMF was prepared according to Organosolv treatment [13]. Briefly, dry powder size of 300 μm from banana peel were alkaline extracted, bleached with hydrogen peroxide and hydrolysed by mixing of nitric acid and acetic acid.

Zein powder from maize with molecular weight of 25,000 - 29,000 was obtained from Fluka. Ethanol (reagent grade, J. T. Baker) and glycerol (Ajax Finechem) were used as-received.

## Nanocomposite Films Preparation

In order to prepared zein/CMF nanocomposite films, zein solution with 16 %wt was prepared by dissolved in 80% ethanol aqueous solution which contain glycerol 20% w/w. The suspension of CMF and zein solution were mixed with 0% - 5% weight fractions of solid CMF in zein matrix. After well mixing of CMF and zein solution, the nanocomposite zein films were casted in a plastic mold and dried at 50°C for 4 hr.

All sample specimens were kept at desecrator with 25°C, 50% RH for at least 48 hr before testing to ensure the stabilization of their moisture content.

# Nanocomposite Film Characterization

Infrared spectra of the as-prepared nanocomposite films were recorded by Perkin-Elmer FTS175 spectrophotometer (FTIR).

TEM morphology of as-prepared CMF was examined by JEOL JEM-1230 while, SEM morphology of as-prepared nanocomposite films were examined JEOL JSM- 6380LV operating at 1.5 kV. Prior observation under SEM, the nanocomposite films were coated with a thin layer of gold.

X-ray diffractometer (XRD) was used for study the crystal structure of the nanocomposite films. XRD patterns were obtained by Bruker axx D8DISCOVER. The X-ray source was Cu K. The measurements covered the scanning range of 10 - 50 at a scanning speed of $0.05 s^{-1}$.

A Mettler Toledo DSC822e (DSC) was used for investigated thermal behavior of the films. To set the thermal history for all samples, each sample was first heated to 200°C and then cooled to 0°C at the scanning rate of 10°C × $min^{-1}$. The thermal properties of the films were measured in the second heating scan at the heating rate of 10°C × $min^{-1}$. The glass transition temperature ($T_g$) and the melting temperature ($T_m$) were determined as the inflection point of the specific heat increment and the onset of the endothermic melting peak of DSC traces, respectively.

Mechanical properties in term of tensile strength, elongation at break and Young's modulus were investigated. These measurements were carried out using a Texture analyzer Stable Micro System TA. XT plus, with the maximum load of 10 kN. A crosshead speed of 10 mm/min and gauge length of 50 mm were used.

# RESULTS AND DISCUSSION

## Morphology

The product of Organosolv treatment from banana peel is presented in the form of colloidal CMF suspension in water media. TEM morphology of CMF was shown in Figure 1. It was found the CMF was fibrous-shaped, long and web-link structure with average diameter of 26 nm.

In manufacturing of protein-based films require plasticization of the protein in order to improve the flexibility of films and make them easy to handle. Then glycerol was added into the zein solution in order to improve zein films flexibility. Figure 2 shows the SEM micrograph of zein films with and without 20% glycerol content. It was found that surface morphology of pure zein film was smooth but fine pores were observed for zein film with 20% glycerol.

The plasticization of a polymer is a complex step-bystep phenomenon, comprising 1) wetting and adsorption; 2) solvation and/or penetration of the surface; 3) absorption, diffusion; 4) dissolution in the amorphous regions; and 5) structure breakdown [18]. Then, the morphology change of zein film, observation of fine pores in the pre- -sented of glycerol was disruption or changing of intermolecular reaction between zein and glycerol [19].

SEM micrographs of zein/glycerol composite films with CMF content of 0 - 5 wt% were showed inFigure 2.

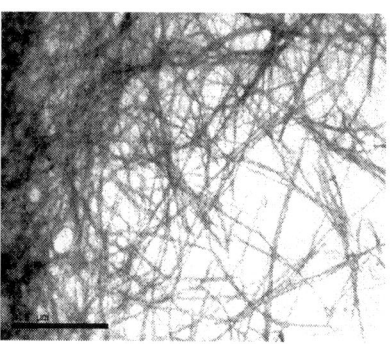

**Figure 1**: TEM micrograph of as-prepared CMF.

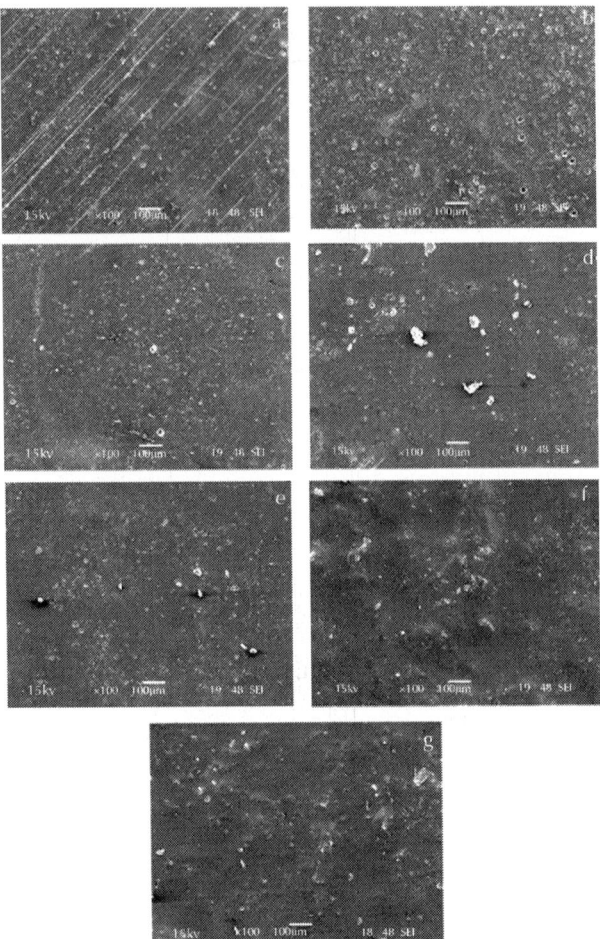

**Figure 2**: SEM micrographs of the surface of zein films having the glycerol content of (a) 0%; and (b) 20%, respectively and zein/CMF nanocomposite films having the CMF content of (c) 1%; (d) 2%; (e) 3%; (f) 4%; and (g) 5%, respectively.

It was found that small white dots were observed and became more clearly with 3% of CMF content. Higher content of CMF, agglomeration of CMF was occurred and it is the reason of roughly surface of zein films. In addition, numbers of the fine pore on the surface of the film were decreased by increasing of CMF content. The reduction in the number of fine pore could be a result of interaction between CMF molecules and zein.

# Chemical Characteristics

Zein/CMF nanocomposite films in various CMF content were prepared using ethanol solution as the common solvent, respectively. To investigate the chemical characteristics of the films, FT-IR was employed. Figure 3 shows FT-IR spectra of pure CMF film (i.e. the topmost curve), and a series of zein/CMF nanocomposite films in various CMF content.

It is apparent from Figure 3 that as-prepared CMF FTIR pattern shows the characteristic hydroxyl-stretching

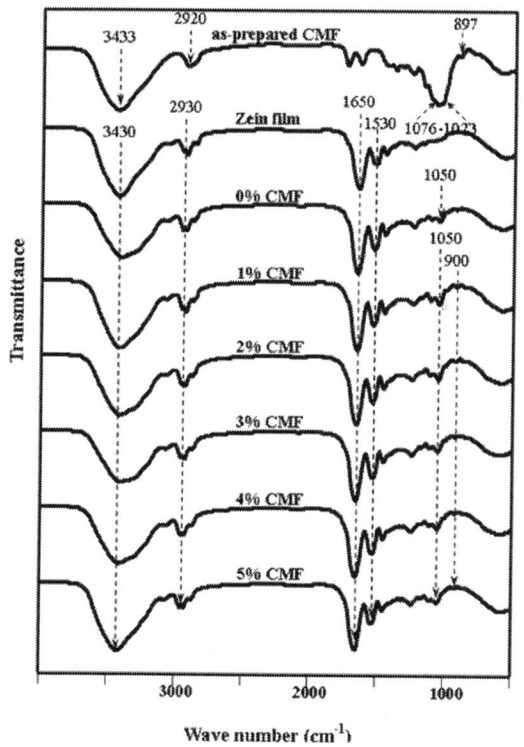

**Figure 3**: FT-IR spectra of as-prepared CMF, and zein/ CMF nanocomposite films.

absorption peak at 3433 cm$^{-1}$, CH stretching absorption peak at 2920 cm$^{-1}$, and characteristic absorption peak of glucopyranose unit of hemicellulose and b-glycosidic linkage between glucose units of cellulose at 1023 - 1017 cm$^{-1}$ and 897 cm$^{-1}$, respectively. The characteristic peaks of zein were appeared at 3430 cm$^{-1}$ (OH stretching), 2930 cm$^{-1}$ (CH stretching) and 1650 cm$^{-1}$ (Amide I and N-H bending), 1530 cm$^{-1}$ (Amide II) which two last groups were related to the type of amino acid in protein molecule [20]. Since the glycerol was added in order to improve the flexibility of zein film, it was found that no significantly shift in characteristic peaks of zein but it was found the appearance of absorption peak at 1050 cm$^{-1}$ (OH bending) of glycerol.

For the zein/CMF nanocomposite films, the FT-IR spectra were found the same characteristic peak as appeared in zein film containing glycerol but the intensity and the broad of these peaks (3430, 2930, 1650 and 1530 cm$^{-1}$) were more pronounce with increasing CMF content. This phenomenon gives absorption peaks at the same position of pyranose group of hemi-cellulose in CMF. Moreover, it was also found the absorption peak at 900 cm$^{-1}$ which refer the b-glycosidic linkage.

Since, apart from the absorption peaks specific to zein/glycerol and CMF is observed in all of the nanocomposite films, no additional peaks signifying possible interaction between zein and CMF were observed. It is logical to postulate that the interaction between zein and CMF may existent but too weak to be detected by FT-IR.

## Crystal Morphology

XRD provides evidence of the crystal morphology. Figure 4 illustrates XRD diffractogram for pure CMF, and a series of zein/CMF nanocomposite films.

As shown in Figure 4, The diffractogram of CMF shows two peaks around 2q = 15.9° with broad shape (peak 1) and 22° with sharp shape (peak 2). This diffractogram was quite altered from the previous report [21] which gave cellulose diffractogram at 2q = 15.5° (peak 1), 16.5° (peak 2), and 22.6° (peak 3). The 2qs were correspond to d values of 5.72, 5.37, and 3.93 angstrom, respectively, and these diffraction peaks are typical of cellulose I. Peak 1 is I (100) and I$_b$ (110); peak 2 is I (010) and I$_b$ (110); and peak 3 is I (110) and I$_b$ (200).

The main peak in the XRD diffractogram of zein films with and without glycerol was appeared at the 2 angle of ca. 18.5°. Since CMF were incorporated into the zein film, it was found no other peak appeared or shifted but the shape of peak at 2 angle of ca. 18.5°

was became broader and shape-like cellulose crystalline peak when increasing CMF content. The incorporation of CMF seemed not to affect the crystallinity of the zein matrices since no relevant changes on their diffraction pattern were observed. This was probably due to the low amount of CMF in the film.

The result from this contribution paper was good agreed with Lu and coworker [22] who study the effect of microfibrillated cellulose on crystal morphology of PVS. It was found that, the diffractograms of the composite PVA films display the superposition of those of the two components. The intensity of the diffraction peaks resulting from microfibrillated cellulose is directly proportional to the concentration of microfibrillated cellulose.

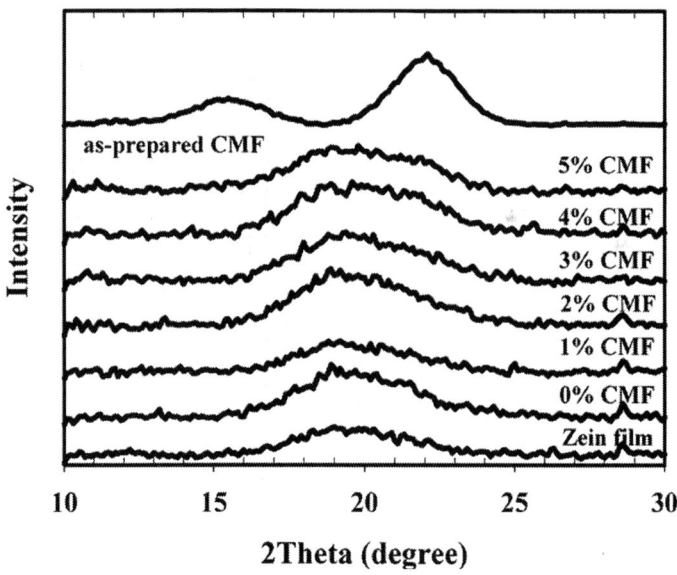

**Figure 4**: XRD pattern of as-prepared CMF and zein/CMF nanocomposite films.

# Tensile Properties

The reinforcement effect of CMF on the mechanical properties of zein/CFM nanocomposite films was evaluated up to their failure, as a function of the CMF content. The tensile strength at break, the percentage of elongation at break, and the Young's modulus of zein/CFM nanocomposite films were shown in Figure 5.

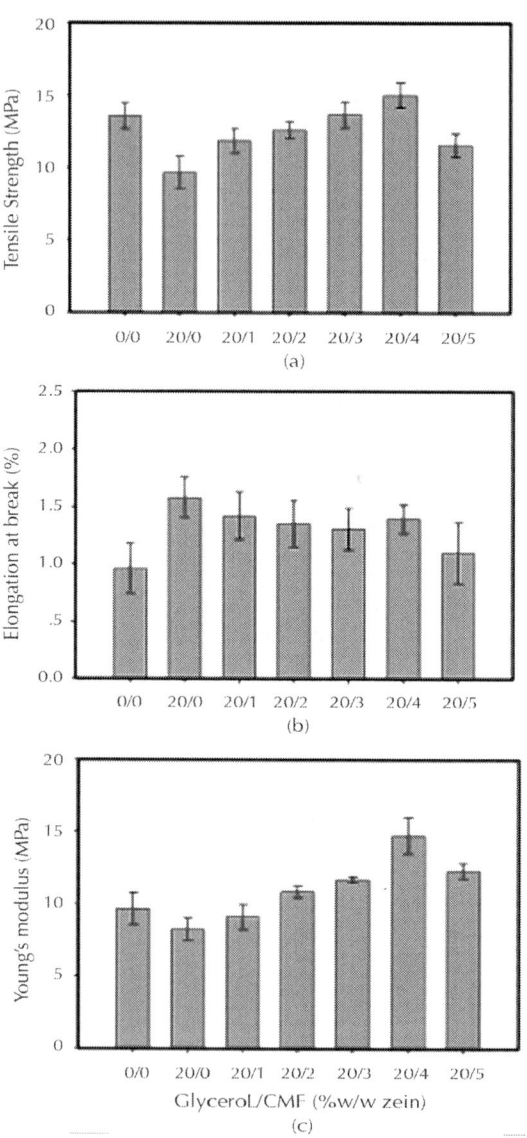

**Figure 5**: Tensile properties of zein/CMF nanocomposite films plotted as a function of glycerol/CMF ratio: (a) Tensile strength at break; (b) Percentage of elongation at break; and (c) Young's modulus.

It was found that pure zein film was brittle with value of 13.58 MPa, 9.97 MPa and 0.96% for tensile strength, Young's modulus and percentage of elongation at break, respectively. In order to improve flexibility of zein film, glycerol was added and it has an improvement the percentage of elongation at break from 0.96 to 1.58. However, glycerol molecules were affecting to the strength and modulus of the film as well. Generally addition of plasticizer such as glycerol or sorbitol is reducing the intramolecular force among polymer chains. It has been attributed to ease of polymer mobility, thus increase in the polymer flexibility. However, there are corresponding lowering the strength and modulus of the polymer chain. These results were similar to report of Parris and Coffin [23] which study the effect of polyethylene glycol on mechanical of zein. It was found that tensile strength of polyethylene glycol added zein film is decreased from 10.9 MPa to 5.7 MPa, while the elongation at break isincreased about by 2.8%.

The reinforcement effect of CMF on mechanical properties of zein nanocomposite was clearly observed. It was found that as CMF content increase from 0 to 5 wt% results in increasing from 11.8 to 15.05 MPa for tensile strength and from 9.11 to 14.75 MPa for Young's modulus of zein nanocomposite films. The highest strength occurs at 4 wt% CMF.

With higher CMF content, aggregation of microfibril reduces both strength and modulus of the films. This result was agree with SEM which shown aggregation of CMF at 5 wt% loading in nanocomposite film. The aggregation of CMF is affected to the strength of zein film by alter the force transfer between zein molecule and CMF. Increasing of CMF content is also affect to slightly decrease the percentage of elongation at break of the composite films. There are two reasons to explain this phenomenon. The first one is higher CMF content caused higher rigidity of composite film. Another explanation is the composite film has higher interaction between cellulose and glycerol, via OH group. It causes lowering the plasticizing effect of glycerol on zein molecules, as a result, the percentage of elongation at break is decreased.

# Thermal Properties

The DSC technique is one of the convenient methods for investigating the thermal properties of polymer blends and composites, therefore it was used for investigation the thermal properties of zein/CMF nanocomposite films on the glass transition ($T_g$) and crystalline melting temperature ($T_m$).

For CMF which is composted of both crystalline part and amorphous part, thus the $T_g$ could not be detected by DSC method. Sun and coworker [24] studied the method to isolation of cellulose from sugarcane bagasse and it was characterized. From the TGA and DSC experiment, it was found that isolated cellulose from nitric acid/acetic acid solution was degraded at 320°C. Especially in DSC thermogram, exothermic peak centered at 340°C was appeared. This exothermic peak is the disintegration of intramolecular interaction and the decomposition of the polymer. However, the degradation study of CMF in this paper was not focus.

Figure 6 shows the DSC thermograms for zein/CMF nanocomposite films. The numbers assigned in the figure are $T_g$ of the nanocomposite films. The thermal transition point of zein is the middle point of endothermic peak at 87°C. This point is correlated to the energy for hydrogen bonding dissociation which related to increasing of protein mobility, thus we postulate that it should be the $T_g$ of zein [25].

There are no baseline shift in the DSC thermogram of biopolymer, the $T_g$ was assumed by the onset, middle or end point of the endothermic peak. As seen in Figure 6, pure zein film gives peak at ca. 81°C but it will be shifted to lower temperature at ca 69.5°C when zein incorporated with glycerol. The lowering of the $T_g$ value of zein by adding glycerol is the plasticizing effect of glycerol

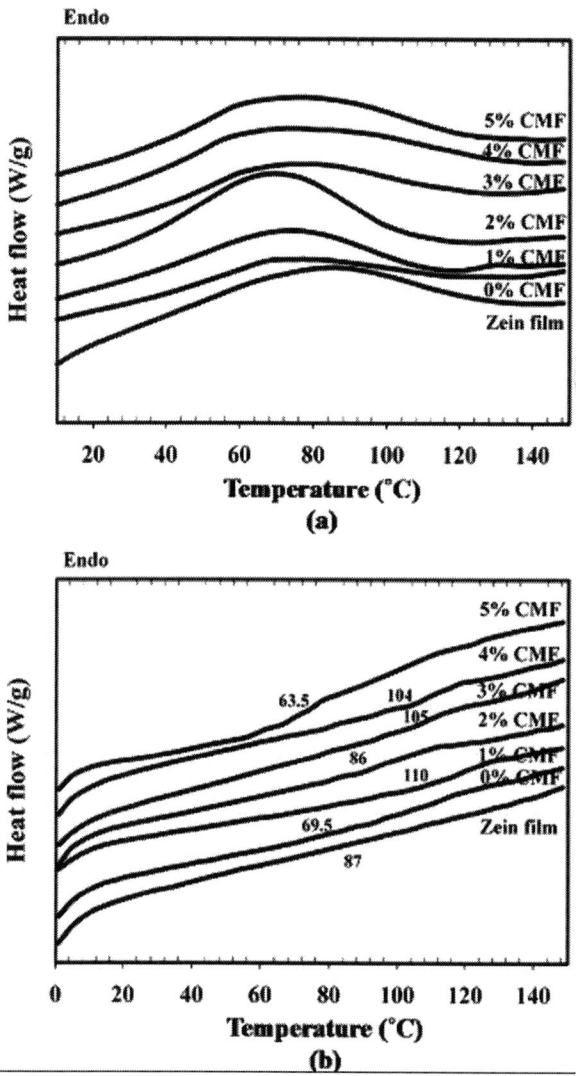

**Figure 6**: DSC thermograms of zein film and zein/CMF nanocomposite films with (a) 1st heating scan; and (b) 2nd heating scan.

which improves the polymer chain mobility. As Ghanbarzadeh and cowork [26] reported that water is the main plasticizer of biopolymer such starch and gluen and result in lowering their $T_g$. The $T_g$ value of anhydrous, high molecular weight starch is 200°C but change to –10°C for hydrated starch. This result is in good agreement to Magoshi and coworker [27] which studying the $T_g$ of zein at different moisture content. It was found that the $T_g$ value of zein is decreased from 139°C to 47°C when increasing moisture content from 0% to 6.6%.

Baseline of DSC thermogram in CMF reinforced zein film is shift to higher temperature as indication of increasing $T_g$ value as increasing of CMF content (Figure 6(b)). This slightly increase in $T_g$ value is probably due to higher restriction chain mobility resulting from the presence of CMF between zein molecules. This result is in good agreement to mechanical properties results which gives higher tensile strength as increasing of CMF content. However as progressive increasing CMF content, the reverse result of $T_g$ value was occurred at 5% CMF content. The decreasing of $T_g$ may results of hydrophilic nature of cellulose, thus an increasing of CMF causes of higher moisture in nanocomposite film at the same testing condition. Though the $T_g$ value of the zein nanocomposite film is decreased with increasing CMF content, the tensile strength is continually increased especially at 4 wt% of CMF. The web-like structure of CMF was the main effect on the increasing tensile strength of the nanocomposite film.

# CONCLUSIONS

Zein/CMF nanocomposite films were developed from colloidal suspension of CMF and zein solution. The CMF, prepared by acid hydrolysis of cellulose from banana peel, consisted of interconnected web-like structure with an average diameter around 26 nm. After the aqueous suspensions of CMF and zein solution were mixed and stirred, solid composite films were obtained by casting and evaporating methods. It was found that an incorporation of CMF was changed the surface morphology and improved tensile properties of the films. Especially, an increasing CMF content was resulted in increasing tensile strength and Young's modulus but decreasing in elongation at break of the zein nanocomposite films. By considering the easy of processing and the mechanical properties, the amount of cellulose microfibrils in zein film should not exceed 4 percent.

# ACKNOWLEDGEMENTS

The support of Department of Tool and Materials Engineering, Faculty of Engineering, King Mongkut's University of Technology Thonburi is gratefully appreciated and acknowledged for the financial support of this research.

# REFERENCES

1.   Y. Li, Y.-W. Mai and L. Ye, "Sisal Fibre and Its Composites: A Review of Recent Developments," Composites Science and Technology, Vol. 60, No. 11, 2000, pp. 2037-2055.doi:10.1016/S0266-3538(00)00101-9

2.   J. Prachayawarakorn, P. Sangnitidej and P. Boonpasith, "Properties of Thermoplastic Rice Starch Composites Reinforced by Cotton Fiber or Low-Density Polyethylene," Carbohydrate Polymers, Vol. 81, No. 2, 2010, pp. 425-433. doi:10.1016/j.carbpol.2010.02.041

3.   H. Ismail, S. Shuhelmy and M. R. Edyham, "The Effects of a Silane Coupling Agent on Curing Characteristics and Mechanical Properties of Bamboo Fibre Filled Natural Rubber Composites," European Polymer Journal, Vol. 38, No. 1, 2002, pp. 39-47. doi:10.1016/S0014-3057(01)00113-6

4.   N. Soykeabkaew, P. Supaphol and R. Rujiravanit, "Preparation and Characterization of Juteand Flax-Reinforced Starch-Based Composite Foams," Carbohydrate Polymers, Vol. 58, No. 1, 2004, pp. 53-63. doi:10.1016/j.carbpol.2004.06.037

5.   T. Nishino, K. Hirao, M. Kotera, K. Nakamae and H. Inagaki, "Kenaf Reinforced Biodegradable Composite," Composites Science and Technology, Vol. 63, No. 9, 2003, pp. 1281-1286. doi:10.1016/S0266-3538(03)00099-X

6.   M. Phiriyawirut, P. Saenpong, S. Chalermboon, R. Sooksakoolrut, N. Pochanajit, L. Vuttikit, A. Thongchai and P. Supaphol, "Isotactic Poly(Propylene)/Wood Sawdust Composite: Effects of Natural Weathering, Water Immersion, and Gamma-Ray Irradiation on Mechanical Properties," Macromolecular Symposia, Vol. 264, No. 1, 2008, pp. 59-66.doi:10.1002/masy.200850410

7.   M. A. S. Azizi Samir, F. Alloin and A. Dufresne, "Review of Recent Research into Cellulosic Whiskers, Their Properties and Their Application in Nanocomposite Field," Biomacromolecules, Vol. 6, No. 2, 2005, pp. 612-626. doi:10.1021/bm0493685

8.   A. Turbak, F. Snyder and K. Sandberg, "Suspensions Containing Microfibrillated Cellulose," US Patent No. 4378381, 1983.

9.   E. Dinand, H. Chanzy and M. R. Vignon, "Suspension of Cellulose Microfibrils from Sugar Beet Pulp," Food Hydrocolloids, Vol. 13, No. 3, 1999, pp. 275-283. doi:10.1016/S0268-005X(98)00084-8

10.  A. Dufresne and M. Vignon, "Improvement of Starch Film Performances Using Cellulose Microfibrils," Macromolecules, Vol. 31, No. 8, 1998, pp. 2693-2696.doi:10.1021/ma971532b

11.  T. Imai, J. L. Putaux and J. Sugiyama, "Geometric Phase Analysis of Lattice Images from Algal Cellulose Microfibrils," Polymer, Vol. 44, No. 6, 2003, pp. 1871-1879.doi:10.1016/S0032-3861(02)00861-3

12.  M. E. Melainine, A. Dufresne, D. Dupeyre, M. Mahrouz, R. Vuong and M. Vignon, "Structure and Morphology of Cladobes and Spines of Opuntia Ficus-Indica. Cellulose Extraction and Characterization," Carbohydrate Polymers, Vol. 51, No. 1, 2003, pp. 77-83. doi:10.1016/S0144-8617(02)00157-1

13.  R. Zuluaga, J. L. Putaux, A. Restrepo, I. Mondragon and P. Ganan, "Cellulose Microfibrils from Banana Farming Residues: Isolation and Characterization," Cellulose, Vol. 14, No. 6, 2007, pp. 585-592. doi:10.1007/s10570-007-9118-z

14.  M. Phiriyawirut, N. Chotirat, S. Phromsiri and I. Lohapaisarn, "Preparation and Properties of Natural RubberCellulose Microfibril Nanocomposite Films," Advanced Materials Research, Vol. 93-94, 2010, pp. 328-331. doi:10.4028/www.scientific.net/AMR.93-94.328

15.  M. Neus Anglès and A. Dufresne, "Plasticized Starch/Tunicin Whiskers Nanocomposites: 1. Structural Analysis," Macromolecules, Vol. 33, No. 22, 2000, pp. 8344-8353. doi:10.1021/ma0008701

16.  J. Sriupayo, P. Supaphol, J. Blackwell and R. Rujiravanit, "Preparation and Characterization of α-Chitin WhiskerReinforced Chitosan Nanocomposite Films with or without Heat Treatment,"

Carbohydrate Polymer, Vol. 62, No. 2, 2005, pp. 130-136. doi:10.1016/j.carbpol.2005.07.013

17.	L. Chazeau, J. Y. Cavaille, G. Canova, R. Dendievel and B. Boutherin, "Viscoelastic Properties of Plasticized PVC Reinforced with Cellulose Whiskers," Journal of Applied Polymer Science, Vol. 71, No. 11, 1999, pp. 1797-1808. doi:10.1002/(SICI)1097-4628(19990314)71:11<1797::AID-APP9>3.0.CO;2-E

18.	J. K. Sears and J. R. Darby, "Mechanism of Plasticizer Action," In: J. K. Sears and J. R. Darby, Eds., The Technology of Plasticizers, Wiley-Interscience, New York, 1982, pp. 35-77.

19.	R. Paramawati, T. Yoshino and S. Isobe, "Effect of Degradable Plasticizer on Tensile and Barrier Properties of Single Plasticized-Zein Film," Journal of Engineering Pertanian, Vol. 1, No. 1, 2003, pp. 49-57.

20.	D. Gioia, L. Guilbert and S. Guilbert, "Corn ProteinBased Thermoplastic Resins: Effect of Some Polar and Amphiphilic Plasticizers," Journal of Agricultural and Food Chemistry, Vol. 47, No. 3, 1999, pp. 1254-1261. doi:10.1021/jf980976j

21.	E. L. Hult, T. Iversen and J. Sugiyama, "Characterization of the Supermolecular Structure of Cellulose in Wood Pulp Fibres," Cellulose, Vol. 10, No. 2, 2003, pp. 103-110. doi:10.1023/A:1024080700873

22.	J. Lu, T. Wang and L. T. Drzal, "Preparation and Properties of Microfibrillated Cellulose Polyvinyl Alcohol Composite Materials," Composites: Part A, Vol. 39, No. 5, 2008, pp. 768-746. doi:10.1016/j.compositesa.2008.02.003

23.	N. Parris and D. R. Coffin, "Composition Factors Affecting the Water Vapor Permeability and Tensile Properties of Hydrophilic Zein Films," Journal of Agricultural and Food Chemistry, Vol. 45, No. 5, 1997, pp. 1596-1599. doi:10.1021/jf960809o

24.	J. X. Sun, X. F. Sun, H. Zhao, R. C. Sun, "Isolation and Characterization of Cellulose from Sugarcane Bagasse," Polymer Degradation and Stability, Vol. 84, No. 2, 2004, pp. 331-339. doi:10.1016/j.polymdegradstab.2004.02.008

25.	F. X. Santosa and G. W. Padua, "Thermal Behavior of Zein Sheet Plasticized with Oleic Acid," Cereal Chemistry, Vol. 77, No. 4, 2000, pp. 459-462.doi:10.1094/CCHEM.2000.77.4.459

26. B. Ghanbarzadeh, A. R. Oromiehie, M. Musavi, Z. E. D. Jomeh, E. R. Rad and J. Milani, "Effect of Plasticizing Sugars on Rheological and Thermal Properties of Zein Resins and Mechanical Properties of Zein Films," Food Research International, Vol. 39, No. 8, 2006, pp. 882-890. doi:10.1016/j.foodres.2006.05.011

27. J. Magoshi, S. Nakamura and K. I. Murakamiki, "Structure and Physical Properties of Seed Proteins, Glass Transition and Crystallization of Zein Protein from Corn," Journal of Applied Polymer Science, Vol. 45, No. 11, 1992, pp. 2043-2048. doi:10.1002/app.1992.070451119

# 10

# Zr-Ti-Ni-Cu Amorphous Brazing Fillers Applied to Brazing Titanium TA2 and Q235 Steel

Jie Cui[1], Qiuya Zhai[1], Jinfeng Xu[1], Yahui Wang[2], and Jianlin Ye[2]

[1]School of Materials Science and Engineering, Xi'an University of Technology, Xi'an, China
[2]Xi'an Unit Container Manufacturing Co., Ltd., Xi'an, China

## ABSTRACT

Ti-Zr-Cu-Ni amorphous filler with good performance is suitable for joining TC and TB titanium alloy, but its melting temperature is higher than 882.5°C, the $\alpha \rightarrow \beta$ phase transition temperature of TA2, which makes the ductility of TA2 fall and the microstructure of the joint course. In this paper, Ti-Zr-Cu-Ni amorphous filler was redesigned and optimized by using orthogonal experiment to obtain three easy-to-use Zr-Ti-Ni-Cu amorphous fillers with low melting points and good plasticity. The fast cooling equipment was used to fabricate the brazing filler foils to implement the braze welding of TA2 and Q235 with high frequency inductance. The results indicate that all the brazing foils are amorphous structure with lower melting temperature, for example, Zr52Ti22Ni18Cu8 filler's is 538°C. The technical parameters in brazing welding are: welding temperature T = 800°C; heating electric current I = 25 A; heating time t = 15 s and holding time t = 15 s, in the case of these conditions, the jointing head shear strength of TA2/Zr52Ti24Ni13Cu11/ Q235 is 139 MPa. Fracture is mainly located in the brazing seam. The white brittle intermetallic TiFe, TiFe2 and enhancement TiC spread in the center zone of brazing seam.

# INTRODUCTION

Since the 1950s, titanium has gradually become an important metal with high specific strength, low density, good thermo stability, tenacity, thermal conductivity and fatigue resistance but high price. Q235 mild steel is a common engineering material with good performance and low price. So, if these two materials can be connected together to be used, their merits can be expressed better, which has good practical worth and economic benefit [1]. However, there is big difference between the physical and chemical properties of titanium and steel, which makes it hard to connect these dissimilar metals. Many methods can be used to connect titanium and its alloy at present [2]. And brazing with simple technology, equipment and low welding temperature is the most appropriate for joining dissimilar metals. Titanium has active chemical property, so it must be brazed under vacuum or dry inert gas atmosphere.

At the moment, the brazing fillers applied to the brazing of titanium and its alloy can be divided into four kinds: Ag-based, Al-based, Pd-based and Ti-based fillers. Through rapid solidification, Ti based fillers can be made into amorphous brazing fillers which has uniform microstructure, little thickness, low welding temperature [3] and good brazing quality et al. Ti-Zr-Cu-Ni alloy is now considered to be the best amorphous filler for brazing titanium alloy, especially in high temperature and severe corrosion environments, but most of this kind of fillers are appropriate for TC and TB series of titanium alloys [4] , rarely for the connection of commercial pure titanium TA2 and mild steel Q235. The melting temperatures of Ti-Zr-Cu-Ni brazing fillers is in a range from 840°C to 900°C lower than the phase inversion temperatures of most titanium alloys, such as the most widely used TC4, whose phase inversion temperature is a range of 980°C - 1000°C [5] [6] ; however higher than the one of TA2, 882.5°C. During the process of heating, when welding temperature is as high as the phase inversion temperature of titanium, α phase transforms into □ phase with obvious coarsening tissue, then becomes acicular α phase during the subsequent cooling process, which makes the plasticity of the base metal TA2 reduced [7] . So it is urgent and hard to acquire a suitable brazing filler for bonding these dissimilar metals TA2 and Q235.

Therefore, the objective of this research is to lower the melting temperature of the Ti-Zr-Cu-Ni brazing filler in order to satisfy the requirement of the welding temperature for brazing TA2 and Q235, and to obtain a brazing filler with good performance appropriate to braze TA2 and Q235. In addition, effects of elements in brazing fillers, performance and microstructure of the fillers and joints will be investigated as well.

# EXPERIMENTAL WORK

Simple metals (99.99%) Ti, Zr, Cu, Ni were melted into alloy by high frequency induction heating equipment in argon atmosphere and brazing fillers were prepared by using a single roller rapid solidification apparatus. The experimental parameters can be seen in Reference [8].

Commercially pure titanium TA2 from Baoji Titanium industry CO. and Q235 mild steel in the form of 50 mm × 10 mm × 3 mm were used in hot rolled and annealed condition for brazing with Zr-Ti-Ni-Cu brazing fillers. The chemical composition of commercially pure titanium TA2 is Fe = 0.25, N = 0.01, O = 0.20, Ti balance. And the chemical composition of Q235 mild steel is C = 0.40, Si = 0.28, Mn = 0.52, P = 0.043, S = 0.040, Ni = 0.30, Cr = 0.29, Cu = 0.28, Fe balance.

The phase structure of the brazing filler was tested by D/MAX-1200 X-ray diffract meter. The melting temperatures of brazing fillers were tested by Netzsch DSC 404C differential scanning calorimetry. The brazing of TA2 and Q235 was conducted by vacuum high frequency brazier, the brazing parameters: vacuum degree is 0.1 Pa; welding temperature is 800°C; heating current is 25 A; heating time is 15 s; holding time is 15 s; cooling to room temperature is in furnace. The sample made along the axis of the welded sample was etched with the solution of 3 mL HF + 6 mL $HNO_3$ + 100 mL $H_2O$, the microstructures of brazing fillers were observed by using Olympus GX-71 metallurgical microscope and JSM -6700F type SEM scanning electron microscope, the shear strength of the joints was tested by WE-100 universal testing machine.

# RESULTS AND DISCUSSION

## Composition of Zr-Ti-Ni-Cu Brazing Fillers

In Ti-Zr-Cu-Ni amorphous brazing fillers, Ni and Cu are stable elements to  phase, which can form eutectic with titanium and reduce the melting temperature significantly [6]. Ni can improve the high temperature property and corrosion resistance of joints [9]. Cu can easily form a lot brittle intermetallic with titanium in joint, so the content of Cu should not be too much. Because of the alloying effect of Zr and Ti, Zr becomes one of the main added elements in Ti-based brazing filler. And Zr can form infinite solid solution with Ti, which can improve strength and keep plasticity. When the content of Zr in the alloy is 50%, the melting temperature of titanium alloy shows a minimum. Zr is neutral in titanium alloy, seldom having effect on the - phase inversion temperature, and it also can form eutectic with Ni and Cu [10]. Therefore, Cu, Ni and Zr are added into Ti-based brazing filler to design ZraTibNicCud, and each element has its content: 48 ≤ a ≤ 60; 20 < b < 28; 3 < d < 12; 19 < c + d < 30; 0.12 < d/(c + d) ≤ 0.5. In order to obtain the Zr-based amorphous brazing filler with good performance, optimized composition of the filler was designed by means of orthogonal experiment.

The orthogonal experiment $L_9$ $(3^4)$ was arranged to search out the optimum brazing filler. In this experiment, the factors are the content of these four elements. According to the approximate content of each element above, three contents of every element were evenly chosen as the level of every factor. So there are 9 experiments with 9 brazing fillers. Indexes are the melting temperature; the tensile strength; the formability and wettability of every brazing foil. Due to the little difference of the tensile strength; the formability and wettability among all the designed fillers, the most important index is the melting temperature. The factors and levels are showing in Table 1.

It is the requirement of this experiment that the melting temperature of the brazing filler should be lower than the phase transition temperature of titanium, 882.5°C, as far as possible. According to this, the most important index, the results from range analysis show that: the dominant factor affecting the melting temperature is the content of Zr, Ti affects less, and Cu affects much less. The compositions of three fillers Zr52Ti22Ni18Cu8, Zr52Ti24Ni16Cu8 and Zr52Ti24Ni13Cu11 are obtained from the largest average combination of every factor's every level. The single effect trends of each element addition on melting temperature are shown in Figure 1.

## Performance of Zr-Ti-Ni-Cu Brazing Fillers

Element Zr has strong glass-forming ability, so the brazing filler containing Zr, Ti, Ni and Cu can easily present amorphous structure. And this kind of researches has been proved a lot. In this experiment the brazing fillers prepared also have amorphous structure and the good performance of this structure. Figure 2 shows the X-ray diffraction spectrum patterns of Zr52Ti24Ni16Cu8 brazing filler. In the pictures there is no peak according to crystal phase, but broad diffraction peaks belong to glassy phases only, which indicate the amorphous structure of Zr52Ti24Ni16Cu8 brazing filler. And all the designed fillers almost have the similar X-ray diffraction spectrums.

The nine brazing fillers with thickness of 40 μm - 60 μm, width of 4mm , prepared by a single roller rapid solidification apparatus, have high plasticity, tensile strength and are convenient to use. The tensile strength of the brazing fillers shows in Table 2. Because in the periodic table, the elements in the fillers are next to the elements in the base metal and they are easily mutually soluble with each other, the wettability of these 9 brazing fillers on base metals is good.

**Table 1:** Orthogonal factor level table

| Level | Factor/at% | | | |
|---|---|---|---|---|
| | Zr | Ti | Cu | Ni |
| 1 | 49 | 22 | 5 | Balance |
| 2 | 52 | 24 | 8 | Balance |
| 3 | 55 | 26 | 11 | balance |

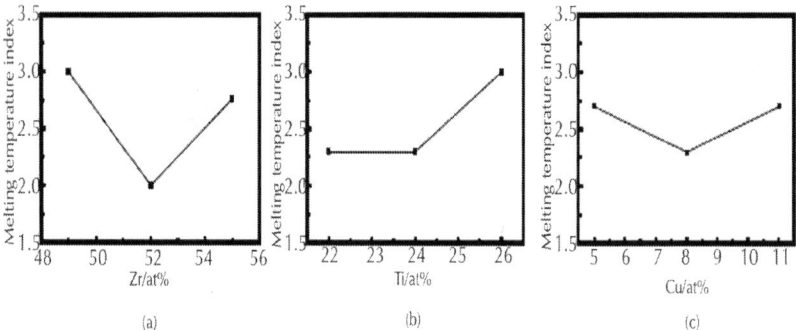

**Figure 1:** Single effect of element addition on melting temperature: (a) Zr; (b) Ti; (c) Cu.

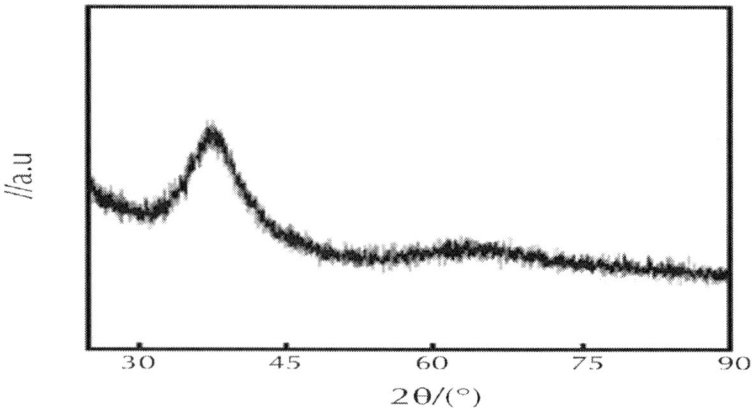

**Figure 2:** XRD spectrum of Zr52Ti24Ni16Cu8 amorphous alloy.

**Table 2:** Strength of extension of the Zr-based fillers

| Brazing alloy | Tensile strength/(MPa) |
|---|---|
| Zr52Ti22Ni18Cu8 | 318.89 |
| Zr52Ti24Ni16Cu8 | 316.84 |
| Zr52Ti24Ni13Cu11 | 307.96 |

Table 3 shows the three Zr-based fillers' melting temperature ranges, it can be seen that the melting temperatures of these three fillers are around 600°C, so the welding temperatures are absolutely under 882.5°C, the α→β phase inversion temperature of TA2, which meets the requirement of this experiment ensuring that the base metal TA2 can keep its fine microstructure and good properties after brazing. The DSC curves of three fillers are showing in Figure 3.

# Microstructure of TA2/Zr-Ti-Ni-Cu/Q235 Brazing Joints

The microstructure of overlap brazing joints is observed. Figure 4(a) shows the whole morphology of the TA2/Zr52Ti22Ni18Cu8/Q235 joint. There appears three zones from top to bottom: base metal TA2 zone, brazing seam zone and base metal Q235 zone. The base metal TA2 near the brazing seam presents saw teeth shape for it partly converts into lath-like structure of   phase. However, the base metal TA2 away from the brazing seam remains the original structure of   phase. Most of the microstructure in the brazing seam zone is dendrite. The light color region in the brazing seam near the base metal Ti is the transition region between the seam and the base metal Ti. There is not an obvious transition region between the seam and the base metal Q235 but a dark color boundary line and the microstructure beside the boundary line is also coarse.

Figure 4(b) shows the microstructure of the overlap joint TA2/Zr52Ti24Ni16Cu8/Q235. For the brazing fillers Zr52Ti24Ni16Cu8 and Zr52Ti22Ni18Cu8 have the almost same compositions, the microstructure of the two joints with the two fillers are very similar. The brazing seam zone of the overlap joint TA2/Zr52Ti24Ni16Cu8/Q235 mainly consists of coarse dendrites. A light color transition region emerges between the seam and the base metal Q235, whose microstructure presents saw teeth shape. There is not an obvious transition region between the seam and the base metal Q235 but a dark color boundary line.

It can be seen in Figure 4(c) showing the microstructure of the overlap joint TA2/Zr52Ti22Ni13Cu11/Q235, that boundaries between the three zones are clear. And between the seam and the base metal TA2, there is also a transition region that consists of light color upper layer and dark color lower layer. There is a narrow transition region between the seam and the base metal Q235, which is not like the joints with the other two brazing fillers, and some white phases emerge on the boundary between the transition region and the seam.

In the seam of the TA2/Zr52Ti24Ni13Cu11/Q235 joint, white region and black region constitute the substrate, on which white dotted phases with different size do not distribute uniformly and small phases gather to form cluster. The center zone of the seam is showing Figure 5. Through spectrum quantitative analysis of point A, it is identified that the white dotted phase is consist of elements C, Ti, Fe and less Zr, Cu, Ni, and the content of Ti is 40.6%, Fe 20.3%, C 28.02%. So the phase contains TiFe2 and TiC compounds.

**Table 3:** Melting range of Zr-Ti-Ni-Cu fillers

| Brazing alloy | Ts/°C | Tl/°C |
|---|---|---|
| Zr52Ti22Ni18Cu8 | 538 | 698 |
| Zr52Ti24Ni16Cu8 | 640 | 741 |
| Zr52Ti24Ni13Cu11 | 613 | 740 |

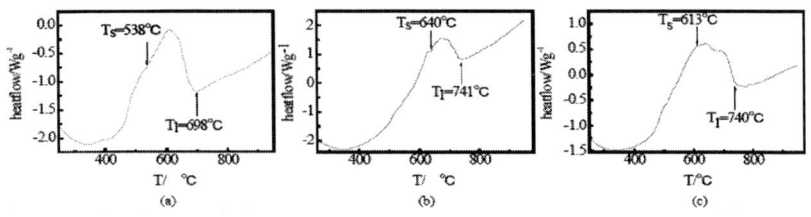

**Figure 3:** DSC curves of brazing ribbons: (a) Zr52Ti22Ni18Cu8; (b) Zr-52Ti24Ni16Cu8; (c) Zr52Ti24Ni13Cu11.

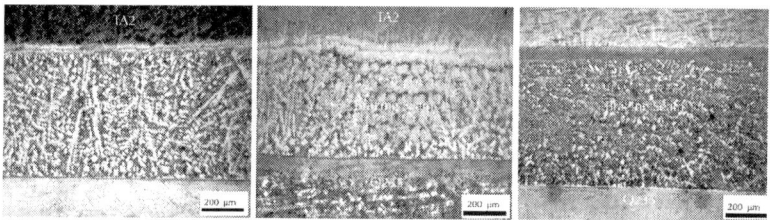

**Figure 4:** The microstructure of TA2/Q235 joints: (a) TA2/Zr52Ti22Ni18Cu8/ Q235; (b) TA2/Zr52Ti24Ni16Cu8/Q235 and (c) TA2/Zr52Ti24Ni13Cu11/ Q235.

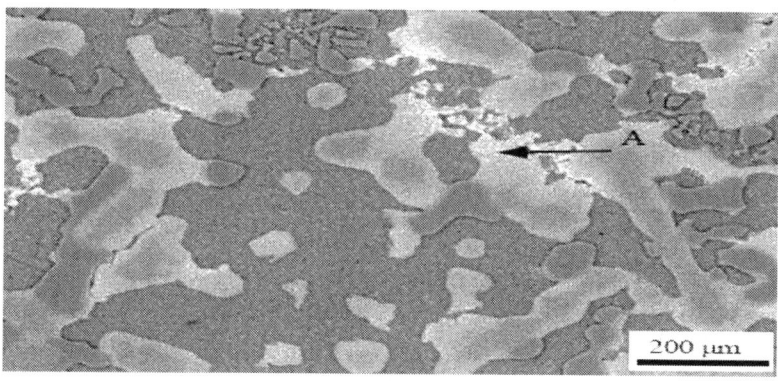

**Figure 5:** Microstructure of TA2/Zr52Ti24Ni13Cu11/Q235.

By analyzing and comparing the microstructure of TA2/ Zr52Ti22Ni18Cu8/Q235, TA2/Zr52Ti24Ni16Cu8/ Q235 and TA2/ Zr52Ti24Ni13Cu11/Q235 joints, it is found that most microstructures of TA2/Zr52Ti22Ni18Cu8/ Q235 and TA2/Zr52Ti24Ni16Cu8/Q235 are coarse dendrites, and no transition region emerges between the seam and the base metal Q235; in TA2/Zr52Ti24Ni13Cu11/Q235 joint, there are white brittle intermetallic compounds $TiFe_2$ and wild phase $TiC$ with different size distributing in the seam, and there is an obvious light color transition region between the seam and the base metal Q235. By comprehensive comparison, the microstructure of TA2/Zr52Ti24Ni13Cu11/Q235 joint is better.

## Mechanical Property of TA2/Zr-Ti-Ni-Cu/Q235 Brazing Joints

The shear strength of the brazing joints was tested, and the highest strength is 139 MPa. From the appearance of facture, it can be seen that the fractures are mainly located in the center of brazing seam.

# CONCLUSIONS

In this paper, Ti-Zr-Cu-Ni amorphous filler was redesigned and optimized by using orthogonal experiment to obtained three Zr-Ti-Ni-Cu amorphous fillers with low melting temperature. The foils were prepared by using a single roller rapid solidification apparatus and high induction frequency brazing of TA2 and Q235 was conducted.

- The brazing foils with amorphous structure have high tensile strength and low melting temperature, under 882.5°C, which meets the requirement of the welding temperature.

- When the brazing parameters are welding temperature T = 800°C, heating current I = 25 A, heating time t = 15 s, holding time t = 15 s, the shear strength of the TA2/Zr52Ti24Ni13Cu11/Q235 joint is 139 MPa. The fractures are mainly located in the seam and white brittle intermetallic TiFe, TiFe2 and reinforced phase TiC spread in the center of the seam.

# ACKNOWLEDGEMENTS

This work was supported by the scientific research project of Shaanxi province science and technology department, the service local special projects of Shaanxi province education department, the integrated innovation plan of Xi'an technology bureau and the western material innovation fund.

# REFERENCES

1. Onzawa, T., Suzumura, A. and Ko, M. (2011) Structure and Mechanical Properties of CP Ti and Ti-6Al-4V Alloy Joints Brazed with Ti-Based Amorphous Filler Metals. Journal of the Japan Welding Society, 5, 205-211.

2. Qi, Y., Zhang, Y.H. and Quan, B.Y. (2003) Development and Application of Braze Welding and Ti-Based Braze Material. Metallic Functional Materials, 10, 31-37.

3. Huang, Y.J., et al. (2008) Formation, Thermal Stability and Mechanical Properties of $Ti_{42.5}Zr_{7.5}Cu_{40}Ni_5Sn_5$ Bulk Metallic Glass. Science in China Series G: Physics, Mechanics and Astronomy, 51, 372-378. http://dx.doi.org/10.1007/s11433-008-0049-y

4. Shapiro, A.E. and Flom, Y.A. (2012) Brazing of Titanium at Temperature below 800°C: Review and Prospective Applications. Welding Journal, 50, 1-22.

5. Chang, H. and Luo, G.Z. (1995) Development of Ti Alloy Used Brazing Filler Metals. Rare Metal Materials and Engineering, 24, 15-20.

6. Zhang, Q.P. and Zhang Y.S. (2005) Technology and the Developmental Situation of the Titanium Alloy. Aerodynamic Missile Journal, 7, 56-64.

7. Takemoto, T. (1988) Intermetallic Compounds Formed during Brazing of Titanium with Aluminum Filler Metals. Journal of Material Science, 6, 1301-1308. http://dx.doi.org/10.1007/BF01154593

8. Xu, J.F. and Wei, B.B. (2004) Liquid Phase Flow and Microstructure Formation during Rapid Solidification. Acta Physica Sonica, 53, 160-166.

9. Elrefaey, A. and Tillman, W. (2007) Interface Characteristics and Mechanical Properties of the Vacuum-Brazed Joint of Titanium-Steel Having a Silver-Based Brazing Alloy. Metallurgical and Materials Transactions, 38, 2956-2961. http://dx.doi.org/10.1007/s11661-007-9357-5

10.  Zhai, Q.Y., Xu, J.F. and Cui, J. (2013) A Kind of Amorphous Brazing Fillers Applied to Brazing Series TA Titanium Alloy and Stainless Steel. The Chinese Patent No. ZL2013104891392.

# Citations

## CHAPTER 1

Hüsnügül Yılmaz Atay and Erdal Çelik, "Electrical Behaviors of Flame Retardant Huntite and Hydromagnesite Reinforced Polymer Composites," ISRN Polymer Science, vol. 2012, Article ID 359034, 9 pages, 2012. doi:10.5402/2012/359034.

## CHAPTER 2

T. LI, S. LI, Y. LI and Z. JIN, "Dechlorination of Trichloroethylene in Groundwater by Nanoscale Bimetallic Fe/Pd Particles," Journal of Water Resource and Protection, Vol. 1 No. 2, 2009, pp. 78-83. doi:10.4236/jwarp.2009.12011.

# CHAPTER 3

Rajani Srinivasan, "Advances in Application of Natural Clay and Its Composites in Removal of Biological, Organic, and Inorganic Contaminants from Drinking Water," Advances in Materials Science and Engineering, vol. 2011, Article ID 872531, 17 pages, 2011. doi:10.1155/2011/872531.

# CHAPTER 4

B. Suresha, G. Chandramohan, J. Prakash, V. Balusamy and K. Sankaranarayanasamy, "The Role of Fillers on Friction and Slide Wear Characteristics in Glass-Epoxy Composite Systems," *Journal of Minerals and Materials Characterization and Engineering*, Vol. 5 No. 1, 2006, pp. 87-101. doi=10.1.1.465.3923.

# CHAPTER 5

Al-Hartomy, O. , Al-Ghamdi, A. , Al-Said, S. , Dishovsky, N. , Ward, M. , Malinova, P. and Mihaylov, M. (2014) A Comparative Study of the Phase Distribution in Carbon-Silica Hybrid Fillers for Rubber Obtained by Different Methods. Materials Sciences and Applications, 5, 685-697. doi: 10.4236/msa.2014.510070.

# CHAPTER 6

Srinivas, K. and Bhagyashekar, M. (2015) Thermal Conductivity Enhancement of Epoxy by Hybrid Particulate Fillers of Graphite and Silicon Carbide *Journal of Minerals and Materials Characterization and Engineering*, 3, 76-84. doi: 10.4236/jmmce.2015.32010.

# CHAPTER 7

K. Devendra and T. Rangaswamy, "Strength Characterization of E-glass Fiber Reinforced Epoxy Composites with Filler Materials," *Journal of Minerals and Materials Characterization and Engineering,* Vol. 1 No. 6, 2013, pp. 353-357. doi: 10.4236/jmmce.2013.16054.

# CHAPTER 8

Cui, J, Zhai, Q, Xu, J, Wang, Y. and Ye, J. (2014) Adding Sn on the Performance of Amorphous Brazing Fillers Applied to Brazing TA2 and Q235. *Journal of Surface Engineered Materials and Advanced Technology,* 4, 342-347. doi: 10.4236/jsemat.2014.46038.

# CHAPTER 9

M. Phiriyawirut and P. Maniaw, "Cellulose Microfibril from Banana Peels as a Nanoreinforcing Fillers for Zein Films,"Open Journal of Polymer Chemistry, Vol. 2 No. 2, 2012, pp. 56-62. doi: 10.4236/ojpchem.2012.22007.

# CHAPTER 10

Cui, J. , Zhai, Q. , Xu, J. , Wang, Y. and Ye, J. (2014) Zr-Ti-Ni-Cu Amorphous Brazing Fillers Applied to Brazing Titanium TA2 and Q235 Steel. *Materials Sciences and Applications,* 5, 823-829. doi: 10.4236/msa.2014.511082.

# Index